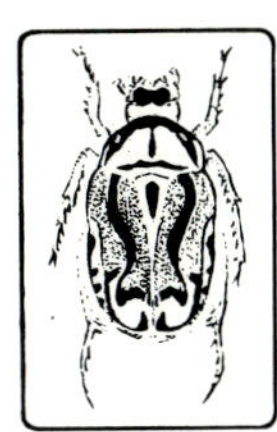

AUSTRALIAN INSECTS IN COLOUR

Frontispiece: Imperial White Butterfly, *Delias harpalyce*

NORLINK
3 0129
ITEM 000 828 906

AUSTRALIAN INSECTS
IN COLOUR

Anthony Healy and Courtenay Smithers

A. H. & A. W. REED

First published 1971

A. H. & A. W. REED

51 Whiting Street, Artarmon, Sydney
357 Little Collins Street, Melbourne
182 Wakefield Street, Wellington
29 Dacre Street, Auckland

ISBN 0 589 07105 X

Photocomposed in Times New Roman by Hartland and Hyde Filmsetting Pty Ltd, Sydney
Printed and bound by Kyodo Printing Company Limited, Tokyo

Plate 1. Noctuid Moth, *Plusia pseudochalcytes*. (Fam. Noctuidae). x6½.

Introduction

THERE ARE PROBABLY more than a million species of animals living now, of which more than three-quarters are insects; thousands more are found and described by entomologists each year. As a group, the insects are the most successful animals that have ever lived. They are found wherever it is possible for a living being to survive; the only habitat they have not really successfully colonised is the ocean. They abound in the tropics and are found at the edge of the polar ice. They are found above the snowline on mountains and they are found deep within caves. They live in the driest deserts and have invaded the freshwaters of streams, ponds, lakes and rivers. It is not only in numbers of species that they surpass all other groups of animals; they occur in enormous numbers of individuals. A single colony of ants may consist of thousands of individuals; a swarm of flying plague locusts may, after landing and feeding for a brief period, leave the area in which it settled devoid of all green vegetation. In some tropical areas thousands of insects may be attracted to a small light; an average-sized orange tree can maintain several thousand insects if it has not been treated with insecticides.

Although insects are small, they make up in numbers what they lack in size. Amongst them are species which feed on our crops and stored food, attack our houses and clothing or spread disease amongst our animals and ourselves. The insects are our main competitors for food in this world of rising populations and food shortages.

From time immemorial some insects have supplied us with food and clothing. The honey-bee and the silkworm have been used by man for centuries. The witchety grub, termites, locusts and other insect species have been articles of diet for man in many parts of the world.

The greatest enemies of insects are other insects; in recent times we have learnt to use some of them in our continuous battle against the pests which attack our crops and possessions.

Quite apart from economic considerations the insects are important to man as objects for study in themselves; the vinegar fly (*Drosophila*) has helped us to reach our present understanding of heredity; the fundamental life processes which go on in all living things, including ourselves, are often more easily studied in insects. The results of such studies can be of practical use in medical, agricultural and veterinary science.

It is important to remember that an insect, as we see it at a particular moment, is at a certain point in the cycle of its development as an individual. In a few species, the youngster has virtually the same appearance and habits as an adult except that it is smaller and is not capable of reproducing. As it feeds and grows it sheds its skin from time to time and eventually becomes sexually mature. This is the simplest form of life cycle, involving mainly increase in size with age leading to maturity. This type of life cycle is found, for example, in the wingless insects known as silverfish.

Most insects undergo greater changes as they develop. The adults usually have wings whereas the immature stages do not. In a considerable number of species the young hatch from the egg in a condition more or less resembling the adult except in size and in the fact that they lack wings. As they feed and grow, they undergo a series of moults, as the silverfish. Sooner or later, however, small outgrowths occur on the back of the thorax and these, the future wings of the adults, become relatively larger at each moult. Eventually, at the final moult, they become functional wings and the insect is able to fly. At this time, or a little later, the insect becomes sexually mature and is able to reproduce. The young in such cases usually feeds on the same food and has habits similar to the adult. This type of development, in which the main changes involve the gradual acquisition of wings, is found in many groups of insects, for example: mayflies, dragonflies, crickets, grasshoppers, stick insects, earwigs, cockroaches, termites, cicadas, leaf-hoppers, aphids, scale insects, bugs and lice. In all of these the young have a similar appearance to their adults and are usually, for convenience, referred to as nymphs.

By far the greater number of species have a third type of life cycle, one which is more complex than either of the previous types. In this, the young which hatch from the egg bear little resemblance to their parents. They are usually different in form, frequently have very different habits, live in different environments and use different food sources. They, too, grow and moult but there is no suggestion of a gradual development of wings. They continue to develop more or less with the form in which they hatched until they reach a point where they no longer need to feed and grow. At this point they usually spin a cocoon or creep away into a protected place and moult once more. The individual which emerges from the cast skin at this moult is very different in form. There is a sudden, tremendous change. It is usually immobile and does not feed but lies quiescent in its cocoon; it has become a pupa (*chrysalis*). Within the pupa the body structures of the individual are reconstructed into adult form and it is from the pupa that the adult insect emerges with wings and sooner or later becomes sexually mature. The young, which are so spectacularly different from the adults, are known as larvae. The appearance of the larvae differs from one group of insects to another and they may be referred to as grubs, caterpillars, maggots or gentles. Insects having this type of life cycle include the lacewings, scorpionflies, butterflies, moths, caddisflies, true flies, fleas, ants, bees, wasps and beetles, as well as many others.

The changes which an insect undergoes during development, however slight, are referred to as its metamorphosis. The advantages of such a type of developmental cycle are many. Having a great difference between larval and adult form enables a species to adapt closely to and exploit two different environments, one as an adult and the other as a larva. As a larva it can specialise in efficient means of obtaining food and we find that larvae are usually adapted to exploit food resources. As an adult it can specialise in reproducing and distributing the species efficiently.

With such a vast number of insects in existence, it is not even possible to mention each group in a book of this nature. It has been necessary to be very selective; the decision on what to put in or leave out has not always been easy but wherever possible the choice has been made with a view to illustrating some general principle which is applicable to a wide range of insects.

The most recent estimate gives over 50,000 species of insects in Australia, of which a very high proportion are not found elsewhere, and in this book we have chosen Australian species to illustrate some of the interesting facets of insect life. Some are bizarre, some beautiful; some are familiar, some are not often seen; some are large and some are small; all form part of a world so far removed from our own that sometimes we can only guess at the significance of their structure or behaviour.

Plate 2. Orchard Swallowtail Butterfly, *Papilio aegeus*. (Fam. Papilionidae) Part of wing pattern, x9.

Acknowledgments

We would like to thank the following for help during the preparation of this book: Brian Bertram for preparing the line drawings; J. Bonwick, J. G. Brooks, Mrs C. Atton, Mrs T. M. Moulds, A. Rose and L. Taylor for providing specimens; Dr D. K. McAlpine, M. Moulds and J. V. Peters for providing specimens, assisting with identifications and reading the text in draft; Dr I. F. B. Common for assistance in identification; Dr H. G. Cogger for providing the photograph of termite mounds; Smila Smithers for assistance in collecting specimens, for preparing the typescript, reading proofs and preparing the index and Dorothy Healy for help during the photography of the live material.

Some Books on Australian Insects

The current major work on Australian insects is *The Insects of Australia*, published by Melbourne University Press, 1970. This is a comprehensive textbook for students and research workers but is not a handbook for identifying species. The butterflies are dealt with in A. N. Burns and E. R. Rotherham *Australian Butterflies in Colour*, a companion volume in the present series and in I. F. B. Common *Australian Butterflies*, Jacaranda Press, 1964. Many moths can be identified from I. F. B. Common *Australian Moths*, Jacaranda Press. Some earlier classic works, such as Froggatt's *Australian Insects*, 1907, Waterhouse's *What Butterfly is That?*, 1932, and Tillyard's *Insects of Australia and New Zealand*, 1926, are out of print and now difficult to obtain; they do, nevertheless, form the basis on which much subsequent work has been done and are still important references for the serious student.

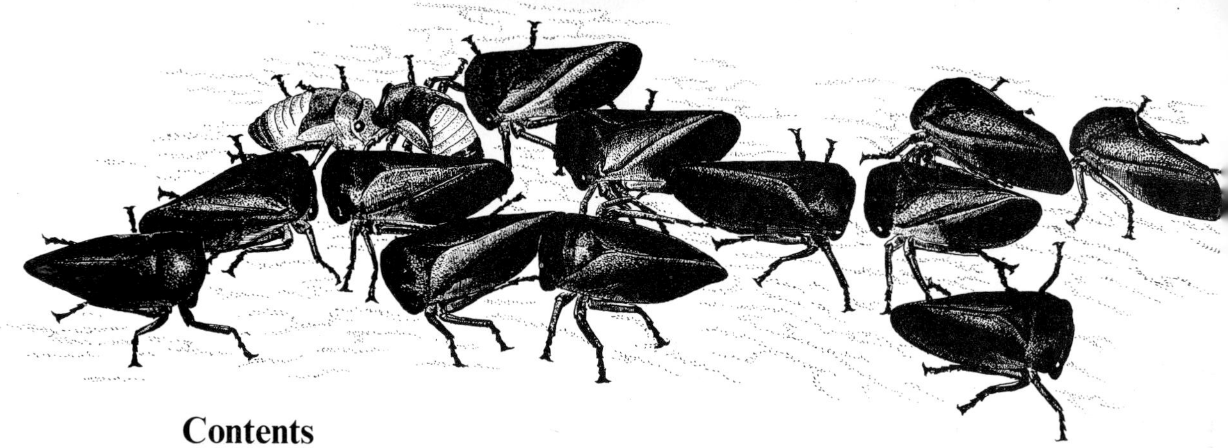

Contents

SILVERFISH
(Order Thysanura)

SILVERFISH (Pl. 3) have become regular inhabitants of Man's dwellings. They are usually found in cupboards and drawers or on bookshelves where they feed on starchy materials such as the glues used in the making of books. In this way they damage bindings or the paper itself; they also make holes in fabrics. They prefer to live in the dark and usually hurry away to cover as soon as they are exposed to light. At no stage in their lives are they winged; they are the descendants of an ancient lineage of insect forms which have never been able to fly.

In the distant past, when the insects' ancestors were developing along lines which led to the many modern winged forms we know today, one line of evolution continued on in its wingless condition to produce the silverfish and its relatives. We do not know the anatomical details of the animals from which insects evolved, but we do know what general features they must have had and some of these are found in modern silverfish; these features are left over, as it were, from the past.

The long-ago ancestors of silverfish were probably animals with long bodies divided into numerous ring-like segments on each of which was a pair of legs. As time went on, legs near the front of the body, just behind the head, were retained as the functional walking legs whilst those on the other body segments further back became reduced or altered in form and adapted to other functions. The head itself is derived from several segments fused together, their legs being modified to form mouth parts which surround the mouth and now function as three pairs of jaws.

Silverfish of today have the typical adult-insect number of six walking legs, but on the undersides of the other segments we can still see the tiny rudiments of legs, reminders of its many-legged predecessors. The fact that silverfish have many features which were probably characteristic of the earliest insects means that they are of great interest to entomologists concerned with studying the evolution of the insects as a whole. Unfortunately, silverfish are soft-bodied animals and so they are unlikely to be fossilised. Fossils of insects which can be regarded as their ancestors have never been found although some relatives have been found in Baltic amber, which is a form of fossilised resin in which insect remains may be embedded. Although not very old by geological standards the insects in amber were living some thirty to forty million years ago.

Silverfish have a sheen on the body which imparts a metallic lustre. The sheen is due to the abundance of minute scales covering the body. These scales are quite beautiful when seen through a microscope, each is ribbed and delicately sculptured. The optical properties of this type of scale structure result in the light falling on the insect being reflected in such a way as to give a silvery sheen. The domestic silverfish is the best-known member of its group but there are native, non-domestic species, which live under bark and stones or in leaf litter feeding on vegetable matter. Some live in the sheltered environment of ants' and termites' nests. Silverfish may live for serveral years. They have the remarkable habit, in the male, of depositing a spermatophore, which is a packet of sperm. The female picks this up and places it in a special receptacle from which the sperm fertilises her eggs as required.

Plate 3. Silverfish, *Ctenolepisma longicaudata*. (Fam. Lepismatidae). x7.

MAYFLIES
(Order Ephemeroptera)

MAYFLIES (Pl. 4) are delicate, winged insects, wellknown to the fly-fisherman. Many of his lures are modelled on mayflies. They have short, fine antennae (feelers) and their filmy wings supported on a network of fine 'veins' are held up above their backs when at rest. (In insects the so-called 'wing veins' are not veins, but are thickenings which form supporting network for the thin wing membrane.) Mayflies are found in swarms over water or sitting on rocks and vegetation alongside streams, ponds and lakes. They are usually inactive by day, but take flight at about sunset. The swarms usually consist mainly of males into which the female flies for mating after which she descends to lay eggs in the water. Some species produce several thousand small eggs and these may all be laid at once or a few at a time. In a few species the female enters the water to attach her eggs to submerged objects but in most cases they are laid as an egg-mass which breaks up on contact with the water. The eggs then sink individually and become attached to stones or weeds.

The nymphs (Fig. 1) creep about on the aquatic vegetation or can be found under stones or in mud and sand on the bottom. Some live in burrows in the stream banks. Along the sides of the abdomen they have pairs of plate-like or filamentous gills through which oxygen is taken from the water and the end of the abdomen carries two or three long terminal filaments. They feed on vegetation growing in the water. The number of moults necessary before the insect becomes adult is relatively high, up to twenty or more. Mayfly nymphs form an important item of diet for some species of fish.

Fig. 1. Mayfly nymph. x2¾.

When a mayfly nymph is ready to emerge as the adult, it floats to the surface of the water. As soon as it arrives at the surface a slit appears in the back of the thorax and the winged mayfly emerges, casting off the old nymphal skin. Expanding the wings to their full size and flying away takes only a few seconds. The winged adult takes off and flies to a resting place nearby. Then, an event which is unique in the insect world takes place—it moults again. Mayflies are the only insects which moult after they have attained the winged state. After the winged individual has moulted, it is fully mature and ready for mating.

Mayflies may live for a few days but some species live as winged adult individuals for only a few hours although they have spent more than a year as an aquatic nymph. The need for mating swarms is clear; mating must occur soon after emergence for time is short in the life of an adult mayfly and by having thousands of individuals emerging at once and forming swarms the meeting of the sexes is assured.

Plate 4. Mayfly. (Fam. Leptophlebiidae). x6½.

DRAGONFLIES AND DAMSELFLIES
(Order Odonata)

DRAGONFLIES are a familiar sight over streams or ponds on warm summer days (Pl. 5), streaking up and down hawking their insect prey or seeking mates while the delicate damselflies (Pl. 6) flit gently over the water. Many of them have striking colour patterns of blues, reds, yellows or metallic iridescent sheens. They are medium to large-sized insects, the biggest with a wing span of six or seven inches. Their wings are membranous, supported by a complex and beautiful lacework of veins.

These insects show some striking adaptations to their way of life. They have existed for millions of years in forms very similar to those which we now see. They are entirely predacious, capturing and eating insects whilst on the wing.

Their structure has become modified to do these things efficiently. They have two pairs of large wings, well suited for rapid flight and great manoeuvrability. They can twist and turn in flight, changing direction so suddenly that their movements cannot be followed by the eye. This skill in flight is essential for an insect which survives only on prey caught in the air. In order to capture flying prey at speed, eyesight must be good. For this reason, they have enormous eyes (Pl. 6) which occupy most of the head, meeting in the middle on top of the head in many cases. Their antennae, which in insects are usually well developed as organs of touch and scent, are reduced to small, hair-like appendages (Pl. 6); they have little use for refinements in these senses.

The thorax, which bears the wings above and the legs below, has been modified so that the legs have moved forward and come to lie near the underside of the head. The legs are strong and armed with spines. These seize the prey and if it is large, hold it to the mouthparts for eating. Smaller prey is dealt with by the jaws alone.

In the course of insect evolution six segments of the body, each with a pair of appendages equivalent to legs, became incorporated into the head structure. The appendages of three of these segments have become modified into the mouthparts of modern insects so that they now have, in effect, three pairs of jaws. The front pair, called mandibles, are usually short, stout and armed with a toothed cutting and grinding edge, used for biting and chewing. The next pair, which lie behind the mandibles is called the maxillae (singular, maxilla). These are often spiny and are used for holding food; they also bear sensitive taste organs. The third pair of jaws are joined together in the middle to form the 'lower lip', called the labium. The mandibles of dragonflies and damselflies are very strong with a jagged cutting edge.

The whole thorax is large and well developed to accommodate the bulk of muscle used to operate the wings during long periods of flight.

Dragonflies and damselflies are unique in their method of mating. The male has a special complex organ situated at the base of the abdomen. He transfers sperm, which issues from the apex of the abdomen as usual in insects, to this organ whilst he is in flight. The male flies along bringing the end of his body round so that the apex touches the underside of his abdomen near the base. The

Plate 5. Dragonfly, *Aeshna brevistyla*. (Fam. Aeshnidae). x2.

sperm is held here until mating takes place. During mating, which also occurs in flight, the male clasps the female behind the head, holding her with a special pair of claspers at the end of his abdomen. The female, which flies below and behind the male, brings the end of her long body below and forward until its tip reaches the organs in which the sperm is stored, near the base of the male abdomen. The sperm is then transferred to the female. Most of the dragonflies either drop their eggs in water or attach them to plants; the damselflies usually embed theirs in slits in plant tissue. It seems that the females are stimulated to lay eggs by the appearance of the surface over which they are flying. Occasionally they are fooled into attempting to lay eggs in unsuitable places; they have been seen trying to lay on a hot tarmac road and on the shiny bonnets of a row of parked cars!

These insects need fresh water to breed successfully, they are often seen flying in inland Australia many miles from the nearest fresh-water and have been seen in large numbers hundreds of miles out to sea. From time to time large migrations of dragonflies are reported and flights may go on day after day with thousands of specimens flying in the same direction over a wide area. The detailed nature, direction and functions of such flights are not known although they are fairly common occurrences and involve mainly sexually immature individuals.

The nymphs of dragonflies (Pl. 6) and damselflies spend their lives in water. In the case of dragonfly nymphs there is a special chamber formed by the rectum which contains the gills. Water is pumped in and out of the anus so that the gills are continuously bathed in a fresh-water supply from which the insect can gain oxygen for breathing. The nymphs of damselflies have three leaf-like processes protruding from the hind end of the abdomen which function as gills. The rectal chamber of dragonfly nymphs has been put to a second use. By sudden and rapid contraction of the chamber by powerful muscles the water contained in it can be ejected forcibly backwards through the anus. The insect is thus shot forward and can move quite rapidly and escape from an enemy by a jerky forward movement.

Like the adults, the nymphs are predacious. What they eat is determined largely by their size in relation to that of the prey and their particular type of habitat; the prey consists of insects, small fish, tadpoles, worms and similar creatures. The nymphs have a remarkably well developed mobile labium, called the mask, which is armed with movable hooks at the end. The nymph approaches its prey slowly and suddenly shoots out its extendible mask, grabbing the prey and bringing it back within reach of the powerful mandibles. The labium is generally not used by insects in prey capture but in dragonfly and damselfly nymphs we see, as so often in insects, one group of species putting an organ common to all to a special use and modifying it accordingly.

Large species may take several years before they are ready to shed their last nymphal skin. Bridges and other structures in water often carry the cast skins of large numbers of dragonflies which have climbed up, emerged as adults and flown away. Some species take quite a long time to acquire their final adult colour; some of the brightest and most spectacularly elegant species are quite dull on emergence.

Plate 6. **Left:** Damselfly, *Ischnura aurora.* (Fam. Coenagrionidae). x4½. **Top right:** Dragonfly nymph. x3¾. **Bottom right:** Dragonfly head, *Hemicordulia tau.* (Fam. Corduliidae). x 10. 

GRASSHOPPERS, LOCUSTS, CRICKETS
(Order Orthoptera)

IN GENERAL APPEARANCE the insects of this order resemble the wellknown 'grasshopper' (Pls. 7-9). The antennae are usually long and are made up of many segments; their mouthparts are adapted for chewing and they have two pairs of wings, of which the front pair is a little thickened and narrower than the hind pair. The latter are folded fan-wise when at rest, the front wings protecting them. Their most obvious feature is the swollen hind legs; these accommodate powerful muscles which enable the insects to jump. The ability to make sounds is fairly general amongst the Orthoptera and with this ability has developed a sense of hearing. As is the case with most insects the usual method of making a sound is by rubbing two parts of the body together. In long-horned grasshoppers and crickets the two fore wings are rubbed together; in short-horned grasshoppers and locusts the enlarged hind legs are rubbed against the thickened fore wings.

The females of most species have a special structure at the end of the abdomen (Pl. 8). It may be short or long, straight or curved. This is an ovipositor, an organ used for egg-laying. The female may embed her eggs in plant tissue or put them in the ground. The soft-shelled eggs pass down through a channel in the ovipositor. It is not as simple as it looks but is made up of elongated plates which are grooved and flanged so as to form a strong, hollow structure. Some Australian species lay their eggs in the open, often along the margins of a leaf; in these cases the eggs are hard-shelled; they do not have the protection from dessication which those within plant tissues have. When the eggs are laid in the ground they are enclosed in a capsule of special material produced by the female.

LONG-HORNED GRASSHOPPERS—
Family Tettigoniidae

The long-horned grasshoppers (Pl. 8) are often heard but seldom seen; their songs form a strong element in the sounds which are so conspicuous and so typically a part of the Australian bush on sultry summer nights. Typically nocturnal, the long-horned grasshoppers have evolved colours and forms which so resemble their usual daytime resting places that they are not often seen and they are more easily located at night by approaching the source of their sound.

In the long-horned grasshoppers the left fore wing overlaps the right wing, which has a special, more or less circular, area near its base (the 'mirror'). A thickened ridge on the left wing is roughened or toothed (the 'file'). The sound is produced by the file being scraped by the edge of the right wing whilst the mirror and the whole wing, acting as a resonator, amplifies the sound. Only the male is capable of sound production. In some species which have reduced wings and cannot fly, the bases of the wings, with the stridulatory apparatus, are still present as short rudimentary wing stubs.

Plate 7. Left: Short-horned Grasshopper, (Fam. Eumastacidae). Front view of head. x9½. **Right:** Short-horned Grasshopper, *Gastrimargus musicus*. (Fam. Acrididae). x3.

To some the sounds produced by these insects are raucous and unpleasant, strident and harsh, monotonous or irritating, but some find them pleasant and relaxing. The sounds produced are not simple; notes of many frequencies are emitted at once and these not continuously. The sound range is such that many notes are well above the range of human hearing; we are hearing only a part of the sound being made. The sound is not smoothly produced but is emitted in pulses. The pattern of pulses is often more important to the insect than the actual notes produced; they tend to recognise the pulse pattern of their own species as an identification feature. The pulse pattern requires special apparatus for its proper analysis; we cannot distinguish the complexities of the sounds with our ears. The main function of sound production seems to be attraction of females to the resting place of the males. The males do not move about much whilst stridulating, but females are usually much more mobile and will approach stridulating males. Stridulation may also be something of a warning to other males that the area is already occupied.

Fig. 2. Long-horned Grasshopper, *Caedicia olivacea,* front leg showing 'ear'. x6½.

The organs which receive the sound are, in the long-horned grasshoppers, situated on the front legs (Fig. 2). Because our ears are on our heads we may sometimes forget that 'ears' might function equally well elsewhere. The auditory organs consist of a membrane which vibrates in response to the sound waves; behind this are structures which receive the vibrations and transform them into nerve impulses which travel along nerves to the central nervous system. The ears of long-horned grasshoppers can receive a much wider range of sounds than we can; this would be expected as the insects produce a wide range of sounds. Our ears become less responsive to a sound if stimulated continuously by it without change, the long-horned grasshoppers seem to retain their responsiveness after prolonged stimulation. In recent years, with the development of more sophisticated equipment for sound detection and analysis, more is being learned about the importance of sound to insects.

Noisy by night the long-horned grasshopper usually sits quietly by day, well camouflaged by its resemblance to its surroundings. The wings of the

Plate 8. **Top:** Long-horned Grasshopper. (Fam. Tettigoniidae). Female with long ovipositor. x2¾. **Bottom left:** Long-horned Grasshopper, *Caedicia olivacea.* (Fam. Tettigoniidae). Adult reared on red petals. x1⅓. **Middle:** Nymph reared on pink petals. x2¼. **Right:** Adult reared on green leaves. x1½.

leaf-inhabiting species are often leaf-shaped, with the veins in the wings resembling the veins of the leaves; some species have even evolved a pattern of spots or other markings which resemble the blemishes which are made on leaves by fungus disease or insect damage.

The green colours are due to pigments in the body derived from their food. When a pigment of another colour predominates in the food supply, this colour may appear on the insect to a greater or lesser degree. Some species, if fed on red rose petals instead of green leaves, will take on a pinkish tinge (Pl. 8); sometimes this will be quite intense and extend over much of the body. The ability to make colour changes in response to the colours of the environment is clearly an advantage, even if it takes time to make the change.

MOLE CRICKETS–
Family Gryllotalpidae

The mole crickets (Pl. 9) have taken to a habitat completely different from that of the long-horned grasshoppers; they spend most of their time below ground in burrows which they have excavated. They have become modified to suit this environment and differ in structure from their close relatives, the long-horned grasshoppers. They have fairly short antennae, extremely long antennae being a disadvantage below ground; their eyes are reduced in size, their fore legs are remarkably strong and well adapted for digging (Fig. 3). The females do not have a greatly elongated ovipositor.

They feed mainly on underground vegetable material but will feed at the surface; they also feed on worms and other insects. Despite the fact that they live underground, the males stridulate to attract the attention of the females; the song can be heard for a considerable distance from the burrow. These insects will cease stridulation in response to vibration of the ground. They do not remain silent for long once the vibrations have stopped and it is easy to locate a burrow by following up the sound and approaching slowly and cautiously. The female lays a group of eggs in a small chamber at the end of a burrow and the young remain in the parent burrow for some time after hatching.

Fig. 3. Mole cricket, *Gryllotalpa nitidula,* front foot adapted for digging. x15.

Plate 9. Mole Cricket, *Gryllotalpa nitidula*. (Fam. Gryllotalpidae). x3.

STICK INSECTS

(Order Phasmida)

THE STICK INSECTS (Pl. 10) are amongst the masters of camouflage in the insect world. In many groups of insects, there are some species which have gained protection from enemies by evolving a form which blends in with their surroundings to such an extent that would-be predators are deceived. In the stick insects this protective device has been carried to extremes. It is to be expected that it would be a docile, vegetarian group of insects which would achieve this. Many species will feed on a wide variety of plants, whereas others have a more restricted diet. Eucalypts and wattles are frequently used as food and there are many grass eating species in Australia.

Some species may be over a foot long; males are usually smaller than females.

Stick insects have long, thin bodies and their legs are also greatly attenuated. The legs are often held close in to the body and the front pair is frequently held forward close together on either side of the head; they are shaped near the base so that the head fits snugly against them (Pl. 10).

As if the shape were not enough, the coloration also resembles that of the plants on which they feed. Grass dwellers are often brown or green according to the grass colour, leaf dwellers are green and some which occur on bark are dark. In this they parallel the mantids and, like the mantids, they also move with a swaying motion. Even when their populations are quite high they are easily overlooked, so effective is their stick-like appearance as camouflage. Large insects are the ones which are the most likely to be sought out as food by birds, small mammals and reptiles and the camouflage devices of stick insects are designed to deceive such predators. Many species have the habit of falling when disturbed. On reaching the ground they lie completely immobile, sometimes for long periods. If roughly handled and held by a leg they are likely to throw it off. Young specimens are capable of re-growing it but it seldom regains full size. As a result of this effective escape mechanism specimens are often found with deformity of legs or antennae.

Frequently, only the males have wings large enough to function, the females being without wings or having small wing rudiments. In many species males are unknown or very rare and the females are capable of laying fertile eggs without mating. These eggs produce females.

Stick insects have hard and often heavily sculptured or ornamented eggs (Pl. 10) with a conspicuous cap-like structure at one end. They are dropped by the females as she climbs amongst the vegetation. They may take one, two or three seasons to hatch. Some females flick them to some distance with a twitch of the hind end of the body as they are laid. A few Australian species of economic importance occur in such numbers as to defoliate forests and control measures may have to be taken against them.

A very large, flightless species (*Dryococelus australis*) occurred on Lord Howe Island but this helpless animal was rapidly reduced in numbers when rats were introduced onto the island. It has not been seen at all on Lord Howe for many years and is probably extinct there. It was recently rediscovered on Ball's Pyramid, a steep-sided islet about fourteen miles from Lord Howe. This is the only place in which this species is known to survive.

Plate 10. Stick insect, *Didymura violescens*. (Fam. Phasmatidae). **Top left:** Adult female. x1. **Top right:** Eggs. x14. **Bottom:** Adult feeding. x5.

EARWIGS

(Order Dermaptera)

EARWIGS (Pl. 11) are essentially nocturnal insects. During the day they can be found under logs and stones or loose bark or amongst vegetation. Some of the largest species are Australian. Their most characteristic and conspicuous feature is the large pair of pincer-like appendages at the hind end of the body. Despite the size of these organs no one is quite certain what their main function is. They are raised and opened when the earwig is disturbed and look quite dangerous but in most cases they have little strength. Earwigs are usually vegetarians or are omnivorous and are not beligerantly carnivorous. The wings of earwigs are peculiar in that the front pair is very short and broad and hardened like the front wings of a beetle. The hind pair of wings is membranous and folded in a complex manner under the front wings when not in use. They are folded fan-wise and also across the pleats into a neat flat parcel which fits under the protective front wings. Earwigs are not often seen in flight but occasionally swarming takes place.

The origin of the common name is not known. These insects display a strong tendency to creep into crevices when disturbed; they will move about until they find a situation in which as much of the body as possible is in contact with some other object and there they will come to rest. This habit is of value to an animal whose main defence is to flee and hide when danger threatens, staying there until some other stimulus, such as hunger, prompts it into activity again. Laboratory experiments with glass blocks can be devised in which an earwig will run and 'hide' from its enemy between the blocks; this is sufficient to satisfy the earwig's sense of security although the insect is still in full view of its supposed 'predator'. This behaviour may seem ridiculous but it is a very effective method of hiding in the wild and has saved the lives of many earwigs over countless generations. This behaviour pattern is as much a part of an earwig as are his forceps.

Unlike most insects, the female earwig of some species, at least, does take a direct interest in the next generation. She makes a small cavity in the soil and lays a group of eggs in it. She remains with them until they hatch and if they are removed and scattered about nearby she will bring them together again. In some cases, she will remain with the young until they disperse. In this behaviour we see the association of parent and offspring, a simple association which, in its most highly developed form, is an essential feature of the great insect societies which we find in ants, bees, wasps and termites.

Plate 11. Earwig, *Labidura riparia*. (Fam. Labiduridae). x5½.

MANTIDS AND COCKROACHES
(Order Dictyoptera)

MANTIDS—*(Suborder Mantodea)*

ALL MANTIDS (Pl. 12) are carnivorous and most will take any suitably sized insect prey. In the mantids we see organs beautifully adapted to carry out their function smoothly, quietly and rapidly. The whole of the front end of a mantid's body is made for capturing prey (Pl. 12). The thorax is elongated in the front part and the first pair of legs situated well away from the second and third pairs. The front legs are not used for walking as much as the others but are held ready for action. The basal section is long and the next is strong with a spiny edge; it is grooved along one side and the third section fits into this, much like a folding knife blade. The legs are held up, folded, and when suitable prey comes within reach, it is seized by a rapid snatching action; once gripped, there is little hope of escape. Most species are coloured to blend in well with their surroundings. Plant dwellers are green; bark dwellers are mottled grey or brown. The camouflage seems to be as efficient at hiding predator from prey as vice versa. Some which live among leaves rock backwards and forwards, like a leaf moving in the breeze.

A mantid can, because of the position and size of its eyes, judge distance very accurately. This enables it to know whether or not prey is within range. In order to be able to judge distance it is necessary for the object to be seen by two eyes simultaneously, the eyes being set a little apart. The large eyes of a mantid are far apart on the mobile head. Mantids are the subject of many superstitions amongst primitive peoples; this is partly due to the peculiar 'praying' stance taken up by many species. The voraciousness of mantids is such that the females may eat the smaller males during or after mating; larger species will eat smaller and the older stages will eat the younger stages of their own species.

The females lay their eggs in large masses (Pl. 12), each mass being enclosed in frothy, parchment-like material. This is produced in a liquid state; air bubbles are incorporated into the liquid and the eggs are laid in the frothy mass which soon hardens. The form of the egg capsule is constant for each species and its function is to protect the eggs. It may provide suitable conditions of humidity and perhaps insulate the eggs from excessive changes in temperature. It does not protect the eggs from insect attack. Even while the female mantid is busy laying her eggs she will often have in attendance some tiny wasps. These wasps are females, waiting to lay their own eggs within the eggs of the mantid as they are laid. The wasp eggs will hatch, each grub will devour the contents of a mantid egg and, after a period as a pupa, will emerge from the egg capsule as a wasp. A remarkably high proportion of mantid egg cases found in nature yield these wasp parasites instead of mantids although, in most cases, a few of the eggs in each capsule escape attack. Perhaps this degree of inefficiency of the wasps is essential if there is to remain a stock of mantids for future wasps to parasitise.

Plate 12. Bottom left: Mantid, *Archimantis latistyla.* (Fam. Mantidae). Head and forelegs. x2½. **Top left:** Same, egg case. x2. **Right:** Mantid, *Orthodera ministralis.* (Fam. Mantidae). x3.

COCKROACHES—*(Suborder Blattaria)*

It may seem strange that entomologists group together such dissimilar insects as mantids and cockroaches. They are, however, remarkably similar in broad features of bodily structure and most of the apparent differences are associated with the specialised way in which the mantids get their food. Cockroaches are mostly nocturnal insects, they hide away by day and become very active after dark. Apart from the wellknown domestic species Australia has a large number of seldom-seen native species. These are found mainly under stones and logs or rotting wood and a few live in the foliage of trees (Pl. 13). Like most insects which hide away, they are dull coloured; the exceptions are mainly those which are openly active during the day. They prefer damp situations but a few species live in desert areas. Cockroaches have remained vegetarians or find a mixed diet to their taste. Some feed on rotten wood. Like the mantids the cockroaches make a special protective covering for their eggs. In the case of cockroaches it is usually a hard, elongate structure (an ootheca). It may be deposited or carried around by the female, protruding from the hind end of her body. Some species deposit it and cover it up with material from the surroundings. In others the ootheca is retained within the female's body until the young are ready to hatch; in these cases it is a thinner membranous structure.

Domestic cockroaches will invade any area in which there is a food supply; their choice ranges from food stores, bakeries and kitchens to sewers. They are also known to be carriers of some human diseases and cannot be thought of as pleasant companions. To their credit, however, they have been extensively studied as laboratory animals and have been used in many of man's researches into problems in biological science.

TERMITES
(Order Isoptera)

TERMITES (Pls. 13, 14) are essentially tropical but they do occur in some warm temperate countries. In Australia they occur most abundantly in northern Queensland, the Northern Territory and northern Western Australia. Within the termite colony the individuals are of different 'castes' and not all the same in appearance or function. The alates ('kings' and 'queens') (Pl. 13) develop wings and are fertile and are the founders of new colonies. Their 'workers' (Pl. 14) are infertile and never become winged but some are capable of reproduction if the kings and queens die. There may also be 'soldiers' (Pl. 14) whose task is to defend the colony.

Some species have entirely subterranean nests and a few maintain their colonies within wood. In order to reach distant food material they will build earthen tunnels; they live and work only hidden from sunlight and in their tunnels the high humidity necessary to them can be maintained. The nests are sometimes complex in structure, made of soil particles which are removed during tunnelling out the underground chambers. Nests may be twenty feet high, although this is exceptional and not all of the material is from the tunnels. Most are much smaller (Pl. 13) and some are quite inconspicuous.

Plate 13. Top left: Winged termite. (Fam. Termitidae). x5. **Bottom left:** Termite nest. (Photo H. G. Cogger). **Right:** Cockroach, *Balta bicolor*. (Fam. Blatellidae). x5.

The nest material is held together by saliva or other secretions and compacted into a hard consistency in the tunnel walls. Walls of inner galleries are softer and digested wood is incorporated in them. The mound as a whole therefore has a hard outer layer and a somewhat softer inner core.

The nest of the magnetic termites (*Amitermes meridionalis*), a northern Australian species, is flattened from side to side so that it has a narrow and a broad side. They tend to be orientated in the same direction, with the broad sides facing east and west. A few species have arboreal nests, but these always have ground connection through tunnels to at least a small subterranean nest.

The light-sensitive workers are pale and colourless. This is why they cannot stand the direct rays of the sun for any length of time. The winged caste is usually darker and is the only one to leave the nest to fly in bright light. Enormous swarms of mature winged males and females sometimes issue from the nests through special exits opened by the workers. The exodus usually takes place at about dusk and in association with rainy or stormy conditions. The millions of flying termites provide a banquet for insectivorous animals of all kinds. Birds and bats attack them in the air, reptiles, insectivorous insects and spiders wait for them to return to the ground. The flight is short, even if the termite is not taken by a predator. The termites form pairs, either in flight or when they have returned to the ground. They shed their wings which are now an encumbrance, and search for a suitable nesting site. The pair runs along the ground, the male closely following the female, until a small crevice or other suitable entrance to the soil is found.

From this a small tunnel is made and mating eventually takes place. The female then lays eggs, usually only a few in her first year. The food necessary to maintain the pair during this period is obtained from substances stored within their bodies, the reserves having been built up when they were in the parent nest. Degeneration of their now useless wing muscles takes place and the products of this process keep the insects alive and independent of outside food sources. The first young produced are workers and as time goes on the ability of the female to produce eggs improves. Eventually, in some species, several thousands of eggs can be produced in a day if the needs of the colony demand it.

Termites feed entirely on wood or other dead plant material. They may feed on material which is specially prepared by certain individuals in the colony and which is regurgitated and produced on demand. Some specialisation of feeding may be found within a colony; soldiers may be fed only on prepared food, also the young members of the colony and the actively reproducing individuals. In nests of some of the more highly developed species are found so-called 'fungus gardens'. These consist of woody material which has been chewed up and mixed with secretions produced by the termites. They may be quite large, and are housed in special chambers within the nest and on them a special fungus is cultivated. This is cropped for food and is tended carefully to prevent contamination. It seems that the fungus is a source of vitamins as well as providing nitrogen-containing food.

The conditions of humidity and temperature within the nests are held remarkably constant by the activities of the insects themselves.

Plate 14. Termites. (Fam. Termitidae). Soldier *(left)*, worker *(right)*. x13. ⇨

SUCKING BUGS
(Order Hemiptera)

AT SOME POINT in the long evolutionary history of the insects, there arose a group which, instead of being equipped with mouthparts which were adapted for biting and chewing up particles of food, evolved mouthparts which enabled them to pierce and suck up the juices from plant tissue. This they achieved by modification of their mouthparts into a set of fine stylets, grooved along their length in such a way that they would fit together to form channels. The whole apparatus became enclosed in a tube-like sheath.

There is a great array of insect species which has arisen from this ancient group and which constitutes one of the most successful of insect orders, the Hemiptera, commonly known as the sucking bugs (Pls. 15-21). Here we find the cicadas, leaf-hoppers, lerps, aphids, scale insects, mealy bugs, shield bugs, bed bugs, water boatmen and so-called water scorpions as well as a host of species for which there are no common names. They all share the ability to feed on liquid foods which are imbibed through their fine probosces.

If an insect is adapted to feeding on a source of food which is abundant only at certain times or for only short periods it is to its advantage to be capable of reproducing rapidly at the time when the resources are readily available and then lie low when they are scarce. Amongst the sucking bugs we find some of the most rapidly reproducing insects known; some in which enormous populations can develop very rapidly from very few individuals. The little damage done by the individual can be magnified enormously when big populations are involved and this, coupled with the ability to spread disease, can result in vast areas of crops being devastated very quickly if action is not taken to suppress the insect population. A few species of bugs protect themselves when nymphs by living in a mass of froth or in a peculiar tube (Fig. 4).

Fig. 4. Leaf hopper, *Chaetophyes* sp., protective calcareous tube secreted by nymph. x4.

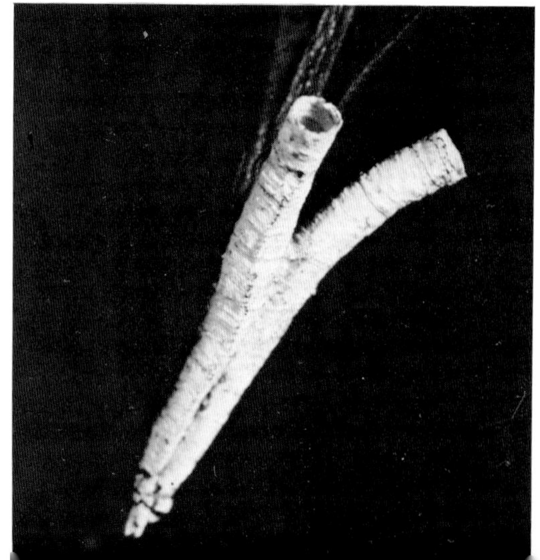

CICADAS—*Family Cicadidae*

Cicadas (Pl. 16) are familiar to most people from childhood, even to city dwellers. Some cicadas are large, handsome and very noisy, attributes which are bound to make them attractive to children. The

Plate 15. Top left: Coreid Bug, *Amorbus rhombifer*. (Fam. Coreidae). x2. **Top right:** Eurybrachid Bug, *Platybrachys decemmaculatus*. (Fam. Eurybrachidae). x6½. **Bottom left:** Flatid Bug, *Siphanta acuta*. (Fam. Flatidae). x9½. **Bottom right:** Shield Bug, *Omyta centrolineata*. (Fam. Pentatomidae). x3.

sounds they make are so much a part of the Australian summer that they have special appeal.

Cicadas lay their eggs in neat rows in slits in branches and twigs. The slits are cut by the female, using a strong ovipositor which is easily seen at the end of her abdomen. Young cicadas, looking very much like a tiny version of their parents, but without wings, drop to the ground on hatching and burrow into the soil. They feed by inserting their mouthparts and sucking out the juices from root tissue. They move through the soil during development and have front legs which are modified for digging. This is analogous to the condition found in mole crickets; insects have solved the problem of subterranean travel many times.

When the cicada reaches maturity it makes its way to the surface, seeks out a vertical support, and after climbing some distance, sheds its skin for the last time to emerge as the fully winged adult cicada. The cast skins are quite wellknown objects in areas where cicadas are emerging as they remain attached to their supports for a while. It may take some time for the newly emerged cicada to attain its full colour and it is quite pale and soft at first; what are to be the darkest areas appear as disrupted dark areas and gradually the colour pattern characteristic of the species appears.

The sounds made by cicadas vary from a very low intermittent hiss to a loud and ear-piercing shrieking which might go on for hours during the heat of the day or even into the night. Only males sing to attract females and when one starts it tends to stimulate others in the area to do likewise; the result is a gradual crescendo of sound when many specimens are present. As is the case with most sound making insects, the song is characteristic for each species and covers a range of sounds much greater than can be heard by the human ear and is more complex than we can appreciate. Some species of cicada are easier to recognise by their song than by looking at specimens and quite clearly they, too, are brought together and recognise members of their own species rather by sound than by sight or scent.

The sound producing mechanism (Pl. 16) in the cicada is on either side of the base of the abdomen. Details of construction vary from simple to complex and sophisticated arrangements. In essence the mechanism is very different from that of most insect sound producers, in which one part of the body is rubbed against another. On each side of the abdomen of the cicada, near the base, is a cavity, which may be covered by an operculum or flap, or may be open. The wall of the chamber is made up of various plates and membranes and behind the whole structure are large internal air sacs. The sound is produced by the rapid in-and-out movement of one of the plates (the tymbal) by the use of special muscles. The mechanism is much the same as that which operates when we push the lid of a tin can in and out to produce a clicking noise, but the speed of this action in the cicada's tymbal is so rapid that the sound produced is a continuous one to our ears. Within the cavity covered by the operculum one of the membranes present is the tympanum, or ear drum of the hearing organs; these are prevented, by special muscles, from being receptive to sounds when the individual is itself making a noise. Amongst the cicadas are the producers of the loudest and most prolonged insect noises, as well as some of the gentlest and most soothing sounds.

Plate 16. Top: Eurymelid Bug, *Eurymeloides pulchra*. (Fam. Eurymelidae). x9. **Bottom left:** Green Grocer Cicada, *Cychlochila australasiae*. (Fam. Cicadidae). x2 **Bottom right:** Double Drummer Cicada, *Thopha saccata*. (Fam. Cicadidae). Stridulatory apparatus. x4.

TREE HOPPERS—*Family Membracidae*

The tree hoppers (Pl. 17) are small insects, remarkable for the bizarre forms which they have evolved. The head is tucked in under the thorax and the latter is produced backward in a long, sometimes strangely shaped, process, which extends to, or beyond, the end of the body. The thorax may also be drawn out into a variety of projections, knobs, horns or other protuberances on either side and these may be ornamented in various ways.

The proportion of water which plant sucking insects take in their food is quite high; hence most of them produce relatively large quantities of sugar-rich liquid excrement. This makes them attractive to other sugar-seeking insects, such as ants. There has developed from this circumstance a situation in which the ants actively seek out the insects producing this so called 'honey dew' and feed upon it. The pugnacity of most ants has given the insects producing the honey-dew some measure of protection from their own enemies because the ants, in the process of attending the honey-dew producers, tend to attack or deter predators. Tree hoppers, sitting in groups on trees and shrubs, are usually attended by ants which stroke them causing them to exude droplets of honey-dew. They are fairly inactive insects but will leap when disturbed. They will sometimes hurry round to hide on the opposite side of branch or twig when approached.

In form and colour many species resemble some feature of the tree species they normally inhabit; in some the projections of the thorax resemble the leaf or bases of leaves and leaf scars which remain when a leaf has fallen. Like cicadas, the tree hoppers deposit their eggs in slits in the bark of the tree, but the young, on hatching, stay on the tree in close groups. *Sextius kurandae* (Pl. 17) occurs widely on wattles in Eastern Australia.

PLANT LICE, GREEN FLY, APHIDS—*Family Aphididae*

Aphids, or plant lice (Pl. 17) are familiar to all gardeners and farmers whose roses, beans or other crops are attacked. Most species spend their lives on the tender young shoots of herbs, trees or shrubs. A few live on roots and a small number of species are responsible for gall formation on plants. Aphids are dependent on the availability of plant hosts in a fresh, growing condition and their life cycle is arranged to take the fullest advantage of temporary food supplies. One species may have several forms during the course of an annual cycle.

When food is abundant and easily obtainable, wingless females occur which are parthenogenetic, that is, they can produce young without mating. The rate of reproduction is also sometimes astonishingly rapid. These females can, in some cases, produce one young a day, and these may start producing their own young at the age of ten days. The young in this case are born alive, that is, the females are viviparous, and their young, too, are all females. A simple calculation will show that this is a tremendous rate of potential increase. By means of this rapid reproductive process the growing shoots of affected plants soon come to carry 'colonies' of aphids.

Within these 'colonies' from time to time, and especially when the shoots become overcrowded, winged females are produced. These winged females take flight and in this way the species spreads to other food plants.

Plate 17. **Top left:** Tree hopper, *Sextius kurandae*. (Fam. Membracidae). x16. **Bottom left:** Rose aphid, *Macrosiphum rosae*. (Fam. Aphididae). Winged adult. x10. **Right:** 'Colony' of rose aphids.

Towards the end of the season, when the quality and quantity of food plant is lowered, these females, as well as the wingless individuals on the original plants, produce males and females. These mate and the female, instead of producing young viviparously, lays eggs. These eggs usually go through the off season, hatching when conditions are again suitable. From the egg comes a wingless, viviparous, parthenogenetic female which starts the cycle again. To take further advantage of what is offering some species feed on one species of plant at one time and on another species at other times so that one species of aphid may be found occupying completely different host plants at different times of the year.

As the colonies develop the aphids are parasitised by minute wasps and a close inspection of a colony of aphids will often show that a high percentage of the aphids are dead and that their bodies have been eaten by the larvae of parasitic wasps.

Like the tree hoppers, aphids produce honey-dew and the leaves of their host plants may become covered by a layer of sugary substance. Not only does this impair the life processes going on in the leaf, but the honey-dew forms an ideal medium on which various species of fungus can grow. The black 'sooty mould' which occurs on fruit trees, especially citrus trees, is a fungus growth which indirectly affects the health of the plant. Aphids are attended by ants for their honey-dew and some species of ants have carried the care which they give to the aphids even further than they have with tree hoppers. In some cases special chambers are built by the ants in the soil in association with roots and the aphids are specially cared for there, being 'milked' of their honey-dew.

SCALE INSECTS—*Family Coccidae and related families.*

The scale insects have, perhaps, the most remarkable appearances of all insects in that the adult females hardly resemble insects at all (Pl. 18); they more usually look like some inanimate attachment or secretion of the plants on which they are found. This form is arrived at by degeneration of many of the more obvious organs during growth and by development of a covering of waxy or powdery secretion.

Scale insects vary greatly in form in the adult state. They may be scale-like, flat, spherical or mussel-shaped. The cottony cushion scale produces a series of wax filaments which give the whole structure a fluted appearance. The antennae and legs of the females are often reduced and nearly always functionless; they are wingless. Scales may be parthenogenic, viviparous or oviparous (laying eggs); one species has been reported as being hermaphrodite, an unusual phenomenon amongst insects.

The young stages of scale insects are more easily recognisable as insects although they are very small. They have short antennae and six functional legs and are known as 'crawlers'. This is the stage of the life cycle which is responsible for spreading the species to new plants and new areas. Owing to their small size, crawlers may be windborne; they are quite hardy and can often live for some time

Plate 18. Top left: Cottony Cushion Scale, *Icerya purchasi.* (Fam. Margarodidae). x10. **Top right:** *Monophlebus* sp. (Fam. Margarodidae). x7½. **Bottom left:** White Wax Scale, *Ceroplastes destructor.* (Fam. Coccidae). x2. **Bottom right:** Gall on *Casuarina* caused by *Cylindrococcus spiniferus.* (Fam. Eriococcidae). x2.

without feeding. After travelling at least some distance from its parent, the crawler settles down and begins to feed. Most of them never move again. As growth proceeds the crawler undergoes a series of moults and legs and antennae are reduced or lost. The reproductive organs develop and by the time the female is mature she is little more than an egg producing organism.

The males remain more mobile than females during their development and in the adult stage are winged. As the female lays eggs she shrivels up and frequently her body, or waxy secretions, come to form a cover for the egg mass. There is a tendency for crawlers not to move very far in some species with the result that large colonies are produced which may be so dense that twigs or branches are covered and eventually, some scale are pushed out and cannot survive. The amount of food removed from a plant by scale insects can be so great as to cause the death of the plant or, in the case of commercially grown plants, to reduce the vitality and yield of the plant or spoil the appearance of fruit. Whilst aphids tend to attack young, fresh, actively growing shoots, scale insects tend to settle and feed on stems and twigs and the longer-lived leaves of evergreen plants.

The peculiar ability to induce gall formation is characteristic of some scale species (Pl. 18); a remarkable number of Australian species have this habit. In these, the crawler, when it settles to feed, causes the plant to start a localised growth which eventually envelopes the scale with plant tissue, known as a 'gall'. The plant responds to a particular species of scale by producing a gall of shape characteristic for that species of scale; in extreme cases, male and females are enveloped in galls of such different shapes, characteristic of the sex of the scale, that they may not be suspected of being produced by the same species. The forms in which galls are found vary from small, simple, structures to large complex abnormalities but they are always of such a structure as to permit the male scale to escape when he is mature and to gain access to the female for mating. The female does not usually leave her gall; the crawlers eventually emerge from the female's gall to disperse.

ASSASSIN BUGS—*Family Reduviidae*

Having evolved an efficient instrument for piercing plant tissues and extracting juices, it is not surprising that another evolutionary step should have been taken and other food sources exploited. In a world so full of insects it is an obvious change for some of the bug species to become predacious and to use other insects as a food supply. This has happened more than once in the history of the Hemiptera and some families have given up the habit of feeding on plants entirely and taken to attacking and feeding on the body juices of insects. This is a nutritious food supply.

Assassin bugs (Pl. 19) are predacious and they have a thickened proboscis which forms a powerful stabbing and sucking organ. Usually, the prey is fairly slow-moving and the active assassin bug merely walks up to the prey and forces its proboscis into a soft area of the body. In other cases the prey may be more agile and the predator has to seize it with its fore legs before jabbing. There is usually remarkably little struggling on the part of a victim attacked by an assassin bug. The prey becomes immobile shortly after it is stabbed.

Plate 19. **Top left:** Assassin Bug, *Poecilosphrodrus wallengreni*. (Fam. Reduviidae). x4½. **Top right:** Toad Bug, *Nerthra alaticollis*. (Fam. Gelastocoridae). x6½. **Bottom:** Bug, *Mutusca brevicornis*. (Fam. Alydidae). x6.

POND SKATERS—*Family Gerridae*

The pond skaters (Pl. 20) have taken to living in a habitat which few animals have succeeded in colonising; they live *on* the surface of still or running water. The surface tension of water is strong relative to the muscular power of insects and to become covered or trapped by a drop of water is a very dangerous situation. Any open body of water, from the water in a knot hole in an old tree stump to that of an open lake is a potential insect trap and large numbers of insects die by falling accidentally into or by landing by mistake on the open surface of water. They become caught in the surface film of the water and, unless they are very strong or can manage to climb onto some floating object, they eventually die either from drowning or from exhaustion.

Pond skaters are long-legged insects which are commonly seen floating, walking or skipping along on the water surface. Some species even venture far out to sea. They have managed to overcome the danger of the water surface by developing hairs which form a pile-like covering over the lower surface of their bodies; this layer prevents the water from wetting the insect and incapacitating it. The whole insect is supported by that very surface film which is such a menace to other species. Their food consists of those less fortunate insects which have been trapped by the surface film of the water; they make use of their supporting medium also as a trap to supply food. Adult pond skaters of many species are winged and they can fly long distances. This they do, like many other aquatic insects, when their pond or stream dries up, when food supplies become short or when the area becomes unsuitable for some other reason. They often fly at night in search of new waters and often alight in swimming pools.

WATER SCORPIONS—*Family Nepidae*

The water scorpions (Pl. 20) are voracious aquatic insects. They live amongst the vegetation, mud and debris at the bottom of streams and ponds. They are long, thin insects and according to size, will feed on almost any water creature. They are able to make long flights from one body of water to another, doing so when the water dries up or becomes unsuitable for some other reason.

Their middle and hind legs are long and slender and are used for creeping about the vegetation. The front legs are adapted for seizing prey and from this the insects get their common name. A large insect living an active underwater life requires a good oxygen supply. At the end of the body are two long appendages each of which has a groove on its inner surface. When placed together they form a tube through which the air supply is maintained. Water scorpions rest suspended from the surface film of the water by the tube or they pierce the surface with the tip of the tube whilst clinging on submerged vegetation. They are quite pugnacious and if carelessly handled will sometimes defend themselves by using their proboscis. The bite can be quite painful and they have become known as 'needle bugs'. Like many other aquatic insects, they frequently appear in swimming pools.

The eggs are laid in a slit which the female cuts in the tissues of a water plant; each egg has a pair of filaments at one end. The young do not have a respiratory tube when they hatch; as they grow the tube develops, becoming relatively longer after each moult.

Plate 20. **Left:** Water Scorpion, *Ranatra australiensis*. (Fam. Nepidae). x2. **Top right:** Water Strider, *Gerris australis*. (Fam. Gerridae). x2. **Bottom right:** Backswimmer, *Enithares woodwardi*. (Fam. Notonectidae). x4.

BACKSWIMMERS—*Family Notonectidae*

The backswimmers (Pls. 20, 21) provide a remarkable example of what happens when a land-dwelling insect evolves into an aquatic one. In many ways the backswimmers resemble other bugs and where they are different, the difference is a clear modification for life in water. They do not differ greatly in general bodily structure from the other bugs but their front legs are reduced in size and the hind legs, instead of being long and thin, are flattened. They swim upside down and the hind legs are fringed with hairs to give added area so that they form very efficient paddles. The body itself has become extremely smooth; ridges and grooves have been eliminated as far as possible so that the body is remarkably sleek and the antennae are small. Although they spend most of their time in water, they must still breathe air; they cannot take oxygen from the water.

In terrestrial insects the air supply is taken in through a series of openings on each side of the body called spiracles. Each spiracle is really the open end of a tube, called a trachea, which branches repeatedly within the body into finer and finer branches. These fine tracheae ramify throughout the insect's body and every organ is served by them. Oxygen is thus brought directly to the tissues along the tracheae and is not taken into lungs and distributed by the blood as it is in human beings. When a backswimmer dives it takes with it a supply of air beneath its wings; it needs to return to the surface from time to time to replenish this supply. Whilst beneath the surface the air in the reservoir is in contact with the tracheal system through the spiracles. Also, the underside of the body has a central ridge with hairs arising from it. Additional ir is trapped below this comb of hairs. Backswimmers e predacious and feed on insects and other aquatic creatures.

SHIELD BUGS—*Family Pentatomidae*

Shield bugs (Pls. 15, 21) are wellknown insects and are recognised by the more or less triangular 'scutellum', a plate which extends along the back from the thorax and overlies the hind margins of the wings when the insect is not flying. The bronze orange bug (*Musgraveia sulciventris*) (Pl. 21) is common on citrus trees. The adults are large and dark brown and feed on the juices of young shoots, causing wilting. The nymphs are oval, extremely flattened and almost disc-like in shape. They are orange or green and are very inconspicuous amongst the foliage of the trees. The eggs are large and brilliant green when laid; the female places them in neat groups, usually on the underside of a leaf.

Most species of shield bugs have the ability to produce a foul-smelling liquid from special glands. The smell is not evident until the insect is disturbed; the liquid then oozes out to the surface or may be squirted out. This ensures wide distribution of the fluid, and acts as a deterrant to further interference.

In a few species the female remains with the eggs after they are laid and may even remain in the vicinity of the young after hatching. The effectiveness of this habit as a protection, however, must be limited as shield bugs have little with which to defend themselves or their young beyond their smell and camouflaging colouration. Their main enemies, as with most insects, are insect parasites which, in this case, attack the eggs. Some shield bug populations are so heavily attacked in this way that it is difficult to find eggs in the field from which to obtain young shield bugs.

Plate 21. Bronze Orange Bug, *Musgraveia sulciventris.* (Fam. Pentatomidae). **Top left;** Eggs, x6 **Top right;** nymph, x3. **Bottom right;** adult, x2 **Bottom left:** Backswimmer, *Enithares woodwardi.* (Fam. Notonectidae). x4.

LACEWINGS, ANTLIONS

(Order Neuroptera)

THE INSECTS of this Order (Pls. 22, 23) are slender-bodied with a delicate appearance, due largely to their wings. These are usually broad, the hind wings being nearly as large as the front pair. All four are membranous with a complex network of fine veins giving them a lace-like appearance. Some of them are extremely beautiful with a pattern of colour on the wings.

They are carnivorous but also feed on honey-dew, are mainly nocturnal and often attracted by light. They fly with rather weak wingbeats and are easily carried by the wind. Although they may play an important role in the general economy of nature, they seldom occur in large numbers. The larvae, which are very different in form from the adults (Pls. 22, 23), are always carnivorous and have their mouthparts modified for extracting the body juices from insects or other prey. The mandibles and part of the maxillae are drawn out and pointed; the mandible is grooved below and into the groove fits the elongated part of the maxilla to form a tube. The combined structures are sickle-shaped and they extend forwards from the head so that those on each side converge towards one another in a pincer-like fashion. The larvae are active and in most cases seek out their prey, impaling them on their pointed mouthparts and sucking out the contents of their bodies. They can sometimes be seen moving around waving their prey impaled on their very efficient mouthparts. The mouthparts are not very strong and the prey is usually soft-bodied. The larvae bear tubercles and hairs on the body surface.

In the uncommon spongilla-flies (family Sisyridae) the larvae are aquatic and feed on the tissues of freshwater sponges. Some species of lacewings are known to congregate in masses in rock crevices; little is known about this habit but these individuals are probably in hibernation.

GREEN LACEWINGS—*Family Chrysopidae*

Some species of green lacewings (Pl. 22) are capable of producing an unpleasant odour from a pair of glands in the thorax when they are handled or otherwise disturbed.

When ready to lay the female exudes a drop of fluid from the end of her abdomen; she touches the substratum on which the egg is to be laid with this and then raises the tip of her body. This draws out the liquid into a thread which rapidly hardens; the egg is laid at the top of this long stalk. Eggs are usually laid in groups if the female is not disturbed. Sometimes the stalks will all arise closely to one another so that a bundle of eggs appears to have been deposited.

The larvae of green lacewings (Pl. 22) are small and spindle-shaped and their main source of food is aphids or scale insects. A larva may eat anything from three hundred to four hundred aphids before being fully developed and ready to turn into a pupa. Thrips, lerps, scale insects and other small, soft-bodied insects are also taken, as well as insect eggs. Green lacewing larvae camouflage themselves by carrying around on their backs the remains of their prey, together with bits of debris from their surroundings. There are hairs on the upper surface of the abdomen with special hooks to which the

Plate 22. **Top:** Green Lacewing, *Chrysopa* sp. (Fam. Chrysopidae). x6. **Bottom left:** Green Lacewing larva, *Chrysopa* sp. x6½. **Bottom right:** Brown Lacewing, *Drepanacra humilis*. (Fam. Hemerobiidae). x10.

remains of the prey are attached. The larva itself is usually whitish, buff coloured or greenish with irregular markings of darker colours such as red, brown or black. When fully mature the larva spins a round silken cocoon in which to pupate. The larvae are strongly cannibalistic and it is possible that the adult habit of placing the eggs on stalks is a habit evolved to overcome the possibility of the first larva to hatch eating the remaining eggs.

ANTLIONS—*Family Myrmeleontidae*

A common and familiar sight in any sandy area are little conical pits in the sand (Fig. 5). These are the traps made by the larvae of the antlion flies (Pl. 23). The adult antlion fly is a fairly large, weakly flying insect with broad, delicate, membranous wings. They fly readily to light. Most species live in dry areas, or inhabit areas where sand is readily available, as in central Australia. They scatter their eggs over the sand and their larvae are responsible for pit formation. The larvae are broad-bodied and have a wide head bearing enormous, sickle-shaped jaws, armed with spines to aid in holding their prey. The larva burrows backwards into the sand and, by means of its head, flicks the loose sand away. By repeating the process in different directions a pit is formed and the larva settles itself below the bottom of the pit, with its jaws protruding above the sand surface.

Fig. 5. Pit traps in sand made by antlion larvae. x2.

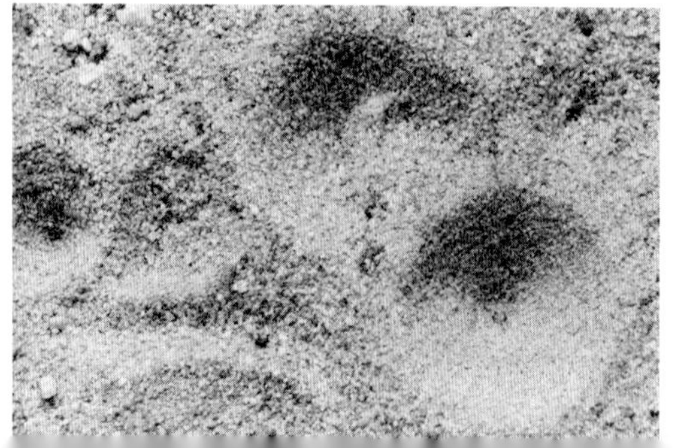

Here it lies in wait for its prey which is frequently, but by no means always, an ant. The prey looses its footing on the steep sides of the pit and falls down to the bottom of the pit; immediately the antlion may be stimulated by the falling sand into flicking this away again. In this way, the scrambling prey tends to be dislodged and brought nearer to the bottom of the pit. Antlion larvae are extremely hardy and can withstand long periods of fasting.

They often travel through the soil just below the surface, by forcing themselves backwards, leaving irregular raised ridges in the sand in areas where they are common. Not all species make pits; some live free in the sand or on tree trunks, others occur under stones or under bark.

Family Myiodactylidae

There is no special common name for the insects of this family, of which there are only a few species known and these occur in Australia, Papua and on Lord Howe Island. Little is known about the adults. They are remarkable for the larval stage, which has a round, flat body with extremely long, fine, curved mouthparts (Pl. 23). The body, around its margin, is extended into processes bearing long hairs and the whole insect is a bright green. It lives on the underside of leaves, especially on plants of the genus *Banksia,* where it feeds on insects. The long jaws are held wide open and snap together when a potential prey walks between them. For a predacious larva it is remarkably quiet and slow moving, relying more on its prey coming to it than on its own ability to seek it out.

Plate 23. **Top left:** Myiodactylid larva, *Osmylops pallida.* (Fam. Myiodactylidae). x7. **Top right:** Antlion larva. (Fam Myrmeleontidae). x3. **Bottom:** Antlion adult, *Glenoleon pulchellus.* (Fam. Myrmeleontidae). x4.

SCORPIONFLIES

(Order Mecoptera)

PRESENT DAY SCORPIONFLIES (Pl. 24) are the modern representatives of a very old order of insects. They have four wings, the hind pair very similar to the front pair, the ends of the wings, which are more or less transparent, are rounded and the head is drawn out into a rostrum with the mouthparts at the end.

They are predacious. The larger species have long legs and when waiting for prey they tend to hang from low herbage by their front legs, allowing the other legs to hang freely. When suitable prey, such as a fly, moth or beetle, comes within range it is seized by the hind legs and carried off to the same or another resting place. It is held in the hind legs which swing it forward to the mouth.

The eggs of scorpionflies may be one of two types. The more primitive species lay oval, smooth eggs in groups in damp soil. The larger, more advanced species lay hard-shelled eggs which are cubical with concave sides (Fig. 6). These are scattered at random and look like grains of sand. The larvae are superficially very much like caterpillars; they feed on dead and dying insects and also on vegetable material on the ground. Some scorpionflies have an interesting courtship procedure. The male captures prey and alights on a shrub or other support and at the same time produces an odour attractive to females. When a female arrives mating takes place, during which the prey is given to the female.

Although the present day scorpionflies are interesting their past is also fascinating. As far back as the Upper Permian Period, some 230 million years ago, there were insects very similar in the essentials of their anatomy as some of the scorpionflies of today. Wings which differ only in details from those of insects now living in the same part of the country have been found in the fossil bearing deposits at Belmont, New South Wales, and in fossil beds in Queensland. These insects have continued to exist, with only relatively minor evolutionary changes, for periods of time which are hard for us to appreciate.

Fig. 6. Scorpionfly eggs, *Harpobittacus tillyardi.* x17.

Plate 24. Scorpionfly, *Harpobittacus tillyardi.* (Fam. Bittacidae). x3. 

MOTHS, BUTTERFLIES
(Order Lepidoptera)

THE MOTHS and butterflies (Pls. 25–34) have always had great popular appeal; there are probably more collectors of butterflies or moths than collectors of almost anything else. This is not surprising because they are very colourful insects and many of them are big enough to be collected and examined without complex or expensive apparatus. They are active and the pleasure of satisfying the hunting instinct is the reward for a hard chase and a successful capture. Also complex metamorphoses which most insects undergo are easily observed in butterflies and moths and they are usually relatively easy to rear in captivity.

The Lepidoptera are a large order of insects and like the other large orders is one in which there remain many species still to be discovered and described; this applies especially to the moths. They have certain features peculiar to them or found uncommonly in other groups. The wings and bodies are clothed in scales (Pl. 25). These are modified hairs, which commonly form a body covering in insects. They are, in fact, flattened and broadened hairs which are usually arranged in such a way as to overlap one another and provide a more or less complete covering to the body, wings and other appendages.

In most insects the colour pattern lies in the integument itself and where the wings are coloured, the wing membranes carry the pigment. With the covering of this by scales, it has become necessary for the scales to carry the colour. The result is that the colour of the scales determines the pattern and by rough handling the scales can be easily removed and the patterns destroyed. The scales appear as fluff or dust when a moth is handled or rubbed. Removal of the scales from the wings does not affect flight provided the wing membrane or the supporting veins are not damaged.

Adult Lepidoptera, that is, the moth or butterfly stage, are usually either nectar feeders or they do not feed at all. The mouthparts are modified to form a proboscis through which nectar is sucked; when not in use, the proboscis is coiled up out of the way under the head. It is soft and flexible and suited to sucking up free liquids. Those with functional mouthparts may have a fairly long feeding period prior to reproduction, sometimes living several months and even hibernating. Those which do not have mouthparts or which have inadequate mouthparts in the adults are usually mature when they leave the pupa and can lay eggs immediately, or they have stored food reserves in their bodies which enable egg development and maturation to take place without feeding. Feeding is also often related to activity; the more active species require food, the more sedentary forms do not use as much energy and so need no additional fuel supply.

The antennae assume a tremendous variety of forms, especially in the males. They may be anything from simple, elongate segmented feelers to large, complex branching organs which are quite beautiful and spectacular. This applies especially to those species in which the male is attracted to the female by scent. The eyes of most butterflies and

Plate 25. **Top left:** Plume Moth. (Fam. Pterophoridae). x6. **Top right:** Forester Moth, *Pollanisis apicalis*. (Fam. Zygaenidae). x5. **Bottom left:** Emperor Gum Moth, *Antheraea eucalypti*. (Fam. Saturniidae). x7. **Bottom right:** Common Jezabel Butterfly, *Delias nigrina*. (Fam. Pieridae). Scale pattern enlarged x50.

moths are large; butterflies can detect small movements of objects in their field of vision and have quite a wide range of colour perception; the eyes of most moths are adapted to make maximum use of a limited amount of light.

Most laying females give the next generation a start by seeking out a plant which is suitable for their caterpillar to feed on. Frequently there is only one species of plant which is suitable and the ability of the females to find a specimen of it is sometimes astonishing. Eggs may be laid singly, or in batches, and they may be laid on selected parts of plants or just scattered far and wide. Moth or butterfly eggs are usually quite beautiful objects, being ridged or with mosaic patterns; they exhibit a variety of shapes and sizes, according to species but tend to be oval, spherical or simple derivatives of these shapes.

The larvae, in the case of moths and butterflies, called caterpillars (Pl 26), are also very varied in colour, shape and size. It is the caterpillar stage of many moth species which are injurious to crops and other products and against which it is often necessary to take active measures. They are mostly of a general cylindrical shape and may be smooth skinned or clothed with hairs to a varying degree, the hairs sometimes arising from fleshy processes on the body. Some are quite furry in appearance. Many are brightly coloured; some have very complex, multicoloured patterns which require magnification before their beauty can be fully appreciated. The caterpillar is, like most larval stages, essentially an animal designed for feeding. When it has fed and grown it pupates either in a cocoon of silk of its own making, or hangs freely from some support (Pl. 26). After a period of internal reorganisation it emerges as the moth or butterfly which is fundamentally the reproductive stage and is also usually responsible for the spread of the species to new areas. Some species may have a fairly short adult life of only a week or two but others may live many months.

SWIFT MOTHS—*Family Hepialidae*

Whilst most caterpillars are vegetarians there are very few which feed on the woody stems of plants. The caterpillars of some of the swift moths are exceptional in that they bore into the wood of living trees, feeding on the wood itself. Others are root eaters. The female swift moth usually distributes her eggs at random by a circular motion of her abdomen as she flies over vegetation or open ground; she lays large numbers of small eggs. This behaviour is in strong contrast to most moths in which the females lay eggs on plants suitable as food for the young caterpillars. Caterpillars of the root feeders tunnel in the soil and feed on the roots of grasses. Like most insects which live in a concealed environment, the caterpillars of swift moths are usually colourless.

The caterpillars, when they are ready to pupate, close the entrance to their tunnel, whether this be in soil or in wood, with a plug of silk, in which other material may be incorporated. The pupa is remarkable in that it moves along its tunnel by means of spines on the abdomen. When the moth is ready to emerge, the pupa moves along the tunnel to the entrance, pushes out the plug made by it when a larva and protrudes itself from the tunnel.

The actual emergence of the moth appears to be

Plate 26. Top left: Caterpillar, *Antheraea eucalypti*. (Fam. Saturniidae). x1½. **Top middle:** Arctiid Caterpillar. (Fam. Arctiidae). x2. **Top right:** Looper Caterpillar. (Fam. Geometridae). x3. **Bottom left:** Cup Moth Cocoon, *Doratifera* sp. (Fam. Limacodidae). x6. **Bottom right:** Caper White Pupa, *Anaphaeis java*. (Fam. Pieridae). x3¾.

strictly determined in some species by weather conditions; in some cases periods of heavy rain induce emergence. This response to an outside influence by many members of the species at once would help to ensure that males and females are active at the same time and so are available for mating. As the adult moths do not feed it is essential for mating to take place before food reserves, built up before emergence, are used up. There are more species of swift moths in Australia than elswhere.

CUP MOTHS—*Family Limacodidae*

One of the most abundantly available sources of nourishing foods for insects is the leaves of plants but they are usually in exposed situations. In order to minimise this disadvantage various protective devices have been evolved by insects which eat them. One of the most efficient is that of the cup moth caterpillar (Pl. 27). These are rather slug-like creatures; the head is hidden below the front end of the body and the underside of the body lies closely applied to the leaf surface. The upper surface is provided with strong irritating hairs and spines arranged in groups and are often folded into pockets hiding them (Pl. 27). When the caterpillar is touched the pockets are everted and the bunches of stinging hairs exposed (Pl. 27). They are very sharp and brittle and on entering the skin the points break off and shatter causing considerable irritation which may continue for days or weeks. Some people are hardly affected, others almost incapacitated.

The efficiency of this type of defence mechanism is obvious and it is most effective against predators which are capable of learning to recognise the offender. Anything which tends to assist this process is an advantage and larvae which are provided with stinging hairs also often have bold colour patterns which make them stand out against their surroundings. The would-be predator is warned well

Fig. 7. White Stemmed Gum Moth caterpillar, *Chelepteryx collesi,* showing irritating hairs. x2.

in time of the nature of the caterpillar and if it has had a previous encounter with one, should heed the warning. Cup moth caterpillars spin a cocoon of tightly packed silken threads which is barrel-shaped or oval and rather parchment-like (Pl. 26). A weak line is left around one end so that when the moth emerges it is pushed up as a lid, permitting the moth to emerge. The moths themselves are small or medium sized and not particularly spectacular.

ANTHELID MOTHS—*Family Anthelidae*

These moths are large or very large, the males having beautifully branched antennae. In some species there is remarkable variation between indi-

Plate 27. Top left: Cup Moth Caterpillar, *Doratifera* sp. (Fam. Limacodidae). Stinging hairs withdrawn. x4. **Top right:** The same, stinging hairs exposed. x4. **Bottom:** White Stemmed Gum Moth Caterpillar, *Chelepteryx collesi.* (Fam. Anthelidae). x2½.

viduals of the same species. They are found only in Australia and New Guinea. The caterpillars are very hairy and some possess irritating hairs. These are often mixed with the other, non-irritating, body hairs (Fig. 7). A very large, common species is *Chelepteryx collesi,* the caterpillars of which (Pl. 27) feed on *Eucalyptus* leaves. When not feeding they rest on the bark of the tree where they are not easily visible. When mature, each spins a somewhat spindle-shaped cocoon and as it does so it forces some of the irritating hairs from its body through the fabric of the cocoon. They retain their irritating properties so that the insect obtains a measure of protection from the larval hairs long after it has become a pupa.

EMPEROR MOTHS—*Family Saturniidae*

The emperor moths, as their common name suggests, are large and spectacular moths, in fact, some of the largest and most beautiful moths in the world belong to this family (Pl. 28). They are often attracted to light and fly, somewhat bat-like, around lamps or into houses. They are strong flyers and have large, heavily built hairy bodies (Pl. 25). The wings are broad and often have conspicuous ring-like markings in some part of the colour pattern (Back cover) and these are exposed if the moth is disturbed when at rest. At the same time the abdomen is bent downwards. This behaviour is clearly intended to discourage any further disturbance from would-be predators.

The caterpillars are large and often brightly coloured with small tufts of hairs arising from tubercles (Pl. 26). They actively feed during the day and it is presumed that they are not often attacked by predators. Despite their size they are not easily found. Caterpillars of this family spin very strong cocoons. In some species the adult moths have a special structure at the base of each front wing which is used for cutting through the cocoon when it is ready to emerge. Some species have larvae which feed in groups and make their cocoons close together. Male emperor moths will fly in to unmated females held in captivity and are probably aware of the female scent for some distance. The most spectacular moth in Australia, the atlas moth (*Coscinocera hercules*) belongs to this family. It is a Queensland and New Guinea species with a wing span of about eight inches. The hind wings are extended backwards into long 'tails' and there are transparent, window-like areas in the wings.

SWALLOWTAILS—*Family Papilionidae*

It is in this family of butterflies that we find some of the most beautiful of all animals (Pls. 28–30). The colours, patterns and beauty of flight of some of the swallowtail butterflies have always attracted, and been admired, by people otherwise completely unfamiliar with insects. Butterfly collectors all over the world are usually keen to add members of this family to their collections and the wings of some species have been used in making jewellery. This has unfortunately led to many of the more beautiful species being considerably reduced in numbers by commercial exploitation. Removal of forests in which they live, however, has probably been the most important single factor in destroying their populations. Regulations are now in force in some countries in an attempt to give protection to threatened species but the only certain way to protect them is to make sure that there are adequate areas of natural forest available to them in which to breed.

Plate 28. **Top:** Emperor Gum Moth, *Antheraea eucalypti*. (Fam. Saturniidae). x1½. **Bottom:** Blue Triangle Butterfly, *Graphium sarpedon*. (Fam. Papilionidae). Eggs, x6¾ **left**; caterpillar, x1¾ **middle**; pupa, x2 **right.**

There are more species in tropical regions than elsewhere but some species do occur in temperate regions. Their common name refers to the slender or tapering extensions to the hind wing which is a feature of many, but by no means all, of the species. The caterpillars of swallowtails are usually very different in their younger stages from what they are just prior to pupation. In the orchard swallowtail (*Papilio aegeus*) the young caterpillar has short, somewhat pointed, protuberances on the body and is dark brownish or greenish-black with irregular white markings. A shiny appearance coupled with its colour pattern and form give it the appearance of a bird dropping adhering to a leaf. The caterpillar feeds and rests openly on the leaf surface. In its later stages (Pl. 30), the predominent colour is green, marked with various shades of brown and cream. The protuberances are relatively small and the insect is well camouflaged by the colour and broken pattern as it sits on a leaf.

In addition to this means of defence against predators, swallowtail caterpillars are capable of producing a pungent odour which, in many of them, is repulsive to human beings; on the other hand some species emit a fairly pleasant odour. The substance giving rise to the odour is produced in special organs, called osmeteria. These are forked processes, often brilliantly coloured and in contrast to the colour pattern of the rest of the caterpillar, which can be extruded out of a pocket behind the head (Pl. 30). They are normally withdrawn but when the caterpillar is disturbed the forked process is everted by blood pressure from within and shoots out in a conspicuous fashion. At the same time the offensive odour is emitted. If a caterpillar is repeatedly stimulated into everting its osmeteria, its reaction becomes slower and slower and it may eventually not react at all. After a period during which there is no stimulation the reaction again becomes rapid. Quite often the odour gives the first indication of the presence of the caterpillar as they are difficult to see and may be disturbed before being detected. The caterpillars are probably distasteful to predators. They have two lines of defence; when the camouflage is ineffectual, they can bring defensive odour into play.

Some species of this family are extremely rare having been seen only a few times; it is thought that these may be hybrids. The blue triangle (*Graphium sarpedon*) (Pls. 28, 29) is, however, a very beautiful common species.

TIGER, CROW AND WANDERER BUTTERFLIES—*Family Nymphalidae*

Subfamily Danainae

The butterflies belonging to this family have their fore legs reduced and functionless; the males of this subfamily have eversible bunches of hairs near the hind end of the abdomen which produce scents for stimulating the females to mate. The wanderer butterfly (*Danaus plexippus*) has the remarkable habit of clustering together in large bunches during the winter in cooler regions. They remain together throughout the winter and may remain at the same site for as long as four or five months. They eventually disperse, the females setting off to find plants on which to lay their eggs. The clusters form on the same trees each year but, of course, the individuals are different each year, being sometimes three or more generations removed from those which occupied the site during the previous winter. The significance of this behaviour is not understood although it must be of some importance in the life of the species.

Most species in this subfamily are quite long-lived, especially in cooler climates and the summer

Plate 29. Blue Triangle Butterfly, *Graphium sarpedon*. (Fam. Papilionidae). x3.

generations usually live for a much shorter time than the winter generations. Whereas many butterflies spend the non-breeding, cold, season in the pupal or even egg stage wanderers spend the winter as adults. During the course of their adult lives the tiger, crow and wanderer butterflies may travel considerable distances in their search for suitable host plants or mates. Sometimes they are found far away from their main breeding areas.

Subfamily Nymphalinae

The nymphs include some very beautiful species, some having fairly simple wing patterns whilst others have quite a complex arrangement of colours. The blue moon or common eggfly (*Hypolimnas bolina*) is a magnificent species in which the male has large blue-white areas on the wings; the species is very variable, however, and there are hardly two females which are exactly the same; it occasionally occurs in New Zealand. An interesting related species, the danaid eggfly (*Hypolimnas misippus*), is usually found north of Brisbane. The male has a generally similar pattern to that of the common eggfly; the female is very different and resembles, in general colour pattern, the lesser wanderer (*Danaus chrysippus*). The lesser wanderer is thought to be distasteful to predators and by having a superficial resemblance to it the danaid eggfly probably has some measure of protection from attack by predators.

Such 'mimicry', in which one species gets protection from enemies by resembling a distasteful species, is found in quite a number of species of butterflies and occurs in many different parts of the world. It also occurs in other groups of insects; there are bugs which mimic ants and there are flies and beetles which mimic wasps Mimicry is carried to a fine degree in some cases, the behaviour of the mimic following that of the species imitated very closely in details.

In the painted lady butterfly (*Vanessa kershawi*) the sexes are very similar to one another. The Australian species is very similar to that of an African and European species and like it, has the habit of undertaking long migratory flights. The Australian painted lady is a common species, its caterpillars feeding on everlasting daisies, cape weed and a few other plants. In spring and early summer in eastern Australia large numbers of the butterfly take to the wing in a persistent flight and set off in a southerly or south-westerly direction. They fly in a set direction, at heights from a few feet above ground to thirty feet or more. They tend to fly over obstacles rather than round them and if they stop for feeding they set off in the same direction as before. This movement may go on simultaneously over a front of several hundred miles and millions of individuals may move at the one time. They can be observed in flight from Queensland to Victoria and the flights may continue for several weeks. There are about thirty other species known to undertake regular migrations in Australia but there are probably many more which have not yet been observed.

Their directions, areas and seasons of migration vary and it is not infrequent to observe different species migrating in opposite directions at the same time and place. The reasons for such large scale movements of populations are not completely understood but such a habit, which is as much a fixed character of the species as its colour pattern, must be beneficial to the species as a whole even though many individuals may be blown to sea or

Plate 30. Top: Orchard Swallowtail Butterfly, *Papilio aegeus*. (Fam. Papilionidae). x1. **Bottom left:** Orchard Swallowtail Caterpillar. x2. **Bottom right:** Caterpillar of Dingy Swallowtail, *Papilio anactus*. (Fam. Papilionidae). x3.

fly out into areas unsuitable for breeding. It is not known whether there is any return movement of the painted ladies; certainly they do not return to the north in massed flight as in their southward movement. It is likely, however, that some subsequently move back.

WHITE BUTTERFLIES—*Family Pieridae*

This family consists of species which are predominently white on the upper side of the wings, although they nearly all have markings in darker colours and many have beautiful striking colour patterns on their under sides (frontispiece, Pl. 31). Many of them are strong migrants, the best studied Australian species being the caper white (*Anaphaeis java*). Observations on the caper white suggest that a southerly movement takes place in very early spring and that this is followed by a later generation making a mass return flight to the north. These northward flights are very spectacular in seasons in which the caper white populations are large and it is in these years that they are usually noticed, although there is some movement each year, even when populations are very low. The main massed movement is in the opposite direction to that of the painted lady and may take place at the same time and place. The plants on which the caterpillars of this species feed are found mainly in the inland areas but the migrations to the north usually bring the butterflies into the coastal areas where food plants (*Capparis* spp.), for their offspring are lacking. It is possible that wind direction plays a part in this. When this happens, the occasional plant in a garden becomes the focal point for many egg-laying females and these plants are seasonally stripped of their leaves by hundreds of caterpillars. The females move on in their migratory flight to lay again further on.

Another common species in this family is the cabbage white (*Pieris rapae*) (Pl. 42). This is a fairly recent addition to the Australian fauna which has established itself virtually wherever cabbages and other vegetables of the same family are grown, as well as in areas where there are suitable wild host plants. Its caterpillars also feed on garden nasturtiums.

When a species gains entry to an area from which it was previously debarred for some reason and it finds the new area satisfactory for development, its population level often rises rapidly. This is because the factors which usually control it in its original habitat are not present in its new area. In the case of insects the major controlling factors are frequently parasitic or predacious insects. The cabbage white has gradually spread through Australia since its arrival in 1937 in Victoria. By the end of 1940 it had spread throughout Victoria and into southern New South Wales, South Australia and Tasmania. Its spread continued and it had reached Western Australia and Queensland by 1943. Suitable climatic conditions and a variety of host plants available for the larvae to feed on, together with the absence of appropriate parasites ensured such a rapid and spectacular spread. By 1943 the cabbage white had become an established part of the Australian fauna. As the cabbage white is an important pest of cabbages and related vegetables, its appearance and spread was of some consequence. Parasitic wasps, known from other areas of the butterfly's range, were introduced and released. These have become established in Australia and now exercise a measure of control on cabbage white populations. One of the parasites attacks the pupal stage of the butterfly, another destroys the caterpillars (Pl. 42).

Several species of this family feed on mistletoe in the larval stage. Using this specialised food source

Plate 31. Caper White Butterfly, *Anaphaeis java.* (Fam. Pieridae). x3.

sometimes restricts the areas in which they are found.

HAWK MOTHS—*Family Sphingidae*

The hawk moths (Pl. 32) are amongst the fastest-flying insects. The body is 'stream-lined', tapering towards the back from a more bluntly rounded front end. The hawk moths have relatively long, narrow, pointed front wings. The hind wings are much smaller than the front and nearly triangular in shape so that their margin forms a straight line with the margin of the front wing. The wings move as one and the speed of movement in full flight is so rapid that the wings are blurred. Speeds of fifteen miles-per-hour and faster have been recorded and can be maintained for long periods. The control of flight is quite remarkable in some species. Sudden darts, interspersed with periods of virtually stationary hovering or even backward movement occur, particularly when the moth is feeding.

Active long-distance, fast-flying species require renewal of fuel supplies and this the moths obtain from the nectar of flowers which open during the evening and night. Their mouthparts are modified to form a long proboscis which, in some species, can be extended to a length as long as, or even longer than, the body. Most moths settle when feeding. Some hawk moths, with their remarkable flying ability, can hover in front of a flower, insert the proboscis and feed, then dart swiftly to the next flower. Species with a long proboscis sometimes 'stand-off' the flower quite a long way when feeding.

The caterpillars of hawk moths (Pl. 32) are usually stout-bodied and smooth, not hairy; nearly all species have a characteristic, obliquely projecting 'horn' near the hind end of the body. In the so-called double-headed caterpillar (*Coequosa triangularis*), which is the larva of the geebung hawk moth, there is no horn-like projection. This species is easily recognised, however, because the last pair of legs at the rear end of the abdomen are greatly enlarged into flap-like claspers above which, on each side of the body, is a large conspicuous dark spot. These have the appearance of eyes and attract attention. The head, of course, is at the other end of the insect, hence the common name for the caterpillar.

A few species of hawk moths have taken to flying by day. Many of these have transparent wings and carry at the hind end of the abdomen a large brush of broad hairs; three species are found in Queensland. In several parts of the world hawk moths undertake long migratory flights. This habit in Australian species has been studied little but with almost sixty species known from this continent it seems very likely that some should be migrants. Some Australian species are known from other parts of the world, suggesting that they are great travellers capable of colonising suitable areas.

Hawk moths have always been favourites with moth collectors. They are spectacular and often brightly coloured; even the more sombre-coloured species have pleasing body and wing patterns. They are often attracted to lights. The caterpillars, when fully developed, leave their host plants and crawl for some distance before finding a suitable place in soil in which to pupate; few species spin a weak silken cocoon amongst leaf litter on the soil surface. Some of the moths have the ability to make a sound. Whereas most sound producing insects rub one

Plate 32. Top left: Oecophorid Moth, *Wingia aurata*. (Fam. Oecophoridae). x5. **Top right:** Hawk Moth, *Coequosa triangularis*. (Fam. Sphingidae). x¾. **Bottom left:** Male of Lasiocampid Moth, *Entometa fervens*. (Fam. Lasiocampidae). x6. **Bottom right:** Caterpillar of Privet Hawk Moth, *Psilogramma menophron*. (Fam. Sphingidae).x1½.

part of the body against another to make the sound, the hawk moths are believed to make their high-pitched sound by forcing air through the proboscis. How this is done is not known but the adults emit an easily audible sound if handled roughly.

ARMY WORMS, CUTWORMS, BOGONG MOTHS, FRUIT PIERCING MOTHS—*Family Noctuidae*

This is the largest family of moths; there are over a thousand species known from Australia. They are usually dull coloured although they may have brightly coloured hind wings; these are usually covered by the front wings when the moth is at rest during the day (Pl. 1). Most of the caterpillars are foliage feeders and many have become major pests. 'Army worms' are noctuid caterpillars which feed on pasture grasses and cereal crops. When they occur in plague proportions they devastate large areas, walking in great hordes and removing plant material as they go. The damage done by such moving bands of caterpillars and the density of their numbers is spectacular and has to be seen to be appreciated. It is believed that the moths which result from these populations also have a pattern of migratory behaviour which leads to their moving into other areas in which subsequent outbreaks of caterpillars occur.

The bogong moth (*Agrotis infusa*) is one of the most interesting species of this family. Whereas most caterpillar species feed actively during the warmer summer months, the bogong moth has evolved a different annual cycle of activity. The caterpillars are 'cut worms' on herbaceous plants which occur in pastures in inland New South Wales and Queensland during the winter. At this time pasture grass areas have a higher weed population than during summer. In early summer the moths resulting from this widespread cutworm population migrate in millions towards the Australian Alps. In years of high population levels they are often reported as being a nuisance in houses or other buildings which they invade temporarily on their journey to the mountains. These migrating individuals have large food reserves stored in their bodies; they pause on their flights to feed at flowering trees and other plants and can often be seen actively swarming around trees in flower throughout the day as well as at night.

When the moths arrive in the mountains they gather together and form compact clusters in rock crevices, usually at heights above 5,000 feet. They remain at these sites until autumn and although both males and females are to be found in the clusters mating does not take place until the clusters break up and the individuals fly back to the pasture areas once again. We see here a clear example of a species which has combined a two-way migration and the ability to enter a period of relative inactivity, with accumulated food reserves, into a cycle which is closely adapted to the appearance and cycle of development of the annual plants on which the caterpillars feed. Most species of insects are adapted closely to fit in their habits and life cycle with the cycle of their food supplies. The evolution of close adaptations to such fluctuating resources is one of the main factors in the success of insects as a group.

Some species of noctuid moths have departed in habits from the general run of the family. In the species of *Catoblemma* the caterpillars have evolved the predacious habit of feeding on scale insects and

Plate 33. Top: Geometer Moth, *Crypsiphona occultaria*. (Fam. Geometridae). x3¼. **Bottom:** Oecophorid Moths, *Zonopetala decisana*. (Fam. Oecophoridae). x6, **left;** *Machimia pudica*. (Fam. Oecophoridae). x3, **right.**

they may be important factors in reducing the numbers of their prey.

In the fruit piercing moths it is the moth stage which has evolved habits somewhat different from the general run. Their caterpillars feed on leaves of various indigenous plants and are found scattered through bushland on their host plants. The moths are capable of piercing and sucking the juices from fruits. They are sometimes the cause of considerable damage and loss of fruit. Although the amount of damage done to each fruit is small, deterioration due to invasion of the wounds by fungi or bacteria is rapid. The fruit becomes marked and unsuitable for sale. A single moth may cause damage to many fruit in this way. It is not easy to control this population and reduction of the pest species by killing the caterpillars is also not easy owing to their being spread through bushland areas beyond orchard limits.

TUSSOCK MOTHS—*Family Lymantriidae*

The tussock moths are inconspicuous insects (Pl. 34). The caterpillars are densely hairy and have very compact tufts of hairs on the upper side. In some species the hairs cause considerable irritation.

In *Orgyia anartoides* (Pl. 34) caterpillars which are to become female moths are much larger than those which give rise to males. This species is common in gardens where the caterpillars feed on a very wide range of garden plants. When it has finished feeding the caterpillar climbs to a fairly high point, on walls, or on the upper branches of a tree, and there spins a rough cocoon of silk. Males eventually emerge and fly off actively. Females, however, are much larger in the body and do not have wings. They merely climb onto the outside of the cocoon and remain there awaiting the arrival of a male. At the time of emergence from the pupa the female already has a large number of fully developed eggs ready for fertilisation and laying. She emits a scent which attracts the males almost immediately and the males are capable of finding the female with remarkable accuracy from a distance, being guided by the scent. The centre of this perception is the male antennae, which are extremely divided and feathery, giving a very large surface area for scent reception. The behaviour in the tussock moths is similar to that of the family Psychidae (bagworms, case moths), in which females are wingless and the males winged. As in the Psychidae, the larvae of tussock moths are responsible for distributing the species as the females are immobile and cannot select host plants on which to deposit their eggs.

Plate 34. Painted Apple Moth, *Orgyia anartoides*. (Fam. Lymantriidae) Caterpillar x5 **top;** Winged male x2¾ and wingless female **bottom.**

FLIES
(Order Diptera)

WHEN THE WORD 'Fly' is mentioned, the immediate thought is of an insect such as the house fly or bush fly. The house fly has, no doubt, been associated with man and his animals for a very long time and the bush fly persists in drawing our attention to itself by its intrusion and numbers at certain times of the year. Related to these flies, are a great many other species of insects, perhaps as many as six thousand species in Australia, some of which are familiar to us and others of which hardly ever come to the notice of anyone but the entomologist who specialises in their study.

Among the Diptera are the crane flies, midges, mosquitoes, sand flies, horse flies, robber flies, soldier flies, fruit flies, blow flies and a host of other groups which have not made themselves wellknown enough for man to have given them common names. Despite great differences in appearance and their diverse habits, all these insects do have certain features in common. They all have only one pair of wings, the front pair, with the hind pair remaining in the form of knobbed structures, known as halteres. The mouth parts of adults are usually adapted in some way for piercing and sucking blood or for lapping up or imbibing fluids; the larvae, which are commonly known as grubs or maggots, do not have any legs and most of them have very reduced heads. As a result of their blood sucking habits the adults of many species have become vectors of diseases which affect man or his animals and medical and veterinary interest in this group of insects has resulted in considerable study of them in relation to disease. Many of the pathogenic organisms which are transmitted by them have a complex association with their vectors and have definite cycles of development in the insect and other hosts affected by them.

Man is still prevented from colonising effectively large areas of the world due to insect-transmitted disease. Most of these diseases are transmitted by insects of this Order. Many species, in the adult stages, are found associated with flowers from which they obtain nectar and with decaying organic materials from which they obtain nourishment or in which their larvae are found. Others feed actively as predators on other insects. A very important group of flies are parasitic, when immature, on other insects. By their predatory and parasitic habits flies as a group, exercise a considerable control on the numbers of other insect species and are probably only second to the parasitic wasps in importance in this respect. Flies form a very important part of our environment and their influence, directly or indirectly, on our well-being, is considerable.

The vinegar flies (*Drosophila* spp.) have been closely studied especially, so far as their genetic systems are concerned, and as a result our knowledge of heredity, and hence of many congenital complaints in man, has been tremendously advanced. Amongst the flies various methods of reproduction have evolved; parthenogenesis is common and so is viviparity. In a small number of species the larvae

Plate 35. **Top:** Fruit Fly, *Euphranta* sp. (Fam. Trypetidae). x17. **Bottom left:** Soldier Fly, *Odontomyia latimacula.* (Fam. Stratiomyidae). x3. **Bottom right:** Robber Fly, *Cerdistus constrictus.* (Fam. Asilidae). x2¼.

are capable of reproduction i.e. maggots give rise to maggots without the appearance of an 'adult' winged stage in the cycle.

CRANE FLIES—*Family Tipulidae*

The crane flies (Pl. 36), also known as 'Daddy-long-legs', are slender with long, thin, easily lost legs and handling a specimen will often break its legs. They include some of Australia's largest flies. They usually have conspicuous antennae and their heads are somewhat elongated. Crane flies are usually encountered amongst shaded, rank vegetation. Their long legs are not well adapted for walking but are used mainly to suspend the insect from vegetation. The larvae are found in water or damp soil where there is an abundance of decaying vegetable matter. Most crane fly larvae are vegetarian but a few are predacious; aquatic predacious species feed on small worms and other water creatures. To those unaccustomed to examining the insects closely, crane flies have a superficial resemblance to mosquitoes, brought about mainly by their having characteristically long, thin legs. They are often mistaken for some kind of giant mosquito; they do not suck blood and mosquitoes do not reach the size of the crane fly species usually mistaken for them. Crane flies are often found in groups, clinging to the undersides of leaves where they bounce up and down on their long legs in a peculiar, rhythmic fashion. Some species have beautifully mottled wings but the majority of species have transparent wing membranes. One of the largest species known (Pl. 36) is Australian.

BEE FLIES—*Family Bombyliidae*

The bee flies (Pl. 36) are usually very hairy and often have stout bee-like bodies. The mouthparts are, in most species, extended into a long thin proboscis but in some it is short. The wings are often dark or bear mottled or bold patterns and they are held in an unusual partly outspread position when the fly is at rest. It is a large family, with more than two thousand species known of which more than four hundred are in Australia. Despite the large number of species and the fact that in some areas at certain times of the year they are fairly common, very little is known about their life histories and general biology. Their rapid, darting flight, with hovering over flowers, and their general form gives them a somewhat bee-like appearance and in some species a remarkable resemblance to wasps has evolved. The flies, of course, have no sting to use if their 'bluff' is called but their mimicry is clearly often successful.

The bee flies are found throughout Australia but they are found in greatest numbers in the hot, dry sandy areas of the interior of the continent. They are parasitic in their larval stages, and they have been reared from bees, caterpillars and flies. In other parts of the world they are often reared from the egg masses of grasshoppers, and locusts, which are deposited in the ground. It seems likely that some Australian species do the same; they are certainly common in areas in which locusts and grasshoppers abound. The females usually deposit their eggs on the soil in areas in which hosts are likely to be found and it is the young fly larva which has to find its way to a host and attack it. Bee flies produce large numbers of eggs, clearly in order to compensate for the losses caused by this uncertain method of host finding.

Plate 36. **Top:** Bee Fly, *Comptosia lateralis.* (Fam. Bombyliidae) x3. **Bottom:** Daddy-long-legs, *Semnotes imperatoria.* (Fam. Tipulidae). Adult, x1 **left;** eggs, x6 **right.**

HOVER FLIES—*Family Syrphidae*

The hover flies (Pl. 37) constitute a widespread family and, often having markings on the body in the form of yellow bands or stripes, are sometimes mistaken for bees or wasps. The drone fly (*Eristalis tenax*) (Pl. 37) which is a stout species of hover fly, frequents daisies and other similar flowers; it has the appearance and flight characteristic of a honey bee. They do look remarkably similar on a flower. The drone fly, of course, is quite harmless; it does does not have a sting. Few people would touch a drone fly on a flower; to most people it is a bee. The maggot of this fly is known as the 'rat-tailed maggot'. It is a large maggot with an extremely long 'tail' at the apex of which are openings to the respiratory system. It inhabits putrid organic matter and is found where decay has progressed to the point of producing a putrid liquid. It extends its 'tail' to the surface to obtain air.

Not all hover fly larvae live in such revolting situations. A few species feed on plants or live in bulbs and these may be present in large enough numbers to become horticultural pests. On the other hand, larvae of hover flies are amongst the most voracious of the predators attacking aphids (Pl. 37). Inspection of aphid infested shoots will often reveal a small white spot. This is a hover fly egg (Pl. 37). The eggs are usually laid near or amongst a group of aphids. The maggot is thus surrounded by its food when it hatches. The adult hover fly feeds on nectar or on the honey-dew which the aphids have dropped onto the leaves.

TACHINID FLIES—*Family Tachinidae*

Tachinid flies (Pl. 38, Fig. 8) are mainly stout-bodied and have an appearance rather like house flies or blow flies. They are nearly all strongly bristled; a few are brightly coloured. As a general principle in insects we find that the parent will make some provision for the survival of the offspring it will never see. The food of tachinid maggots consists of the internal tissues of other insects. The ways in which tachinids ensure that their maggots reach suitable hosts are many, each species being fixed in its method. In the simplest cases the female seeks out the kind of habitat in which the appropriate host is to be found and lays her eggs there. The maggot leaves its egg and goes in search of a host insect. When it finds one, usually a caterpillar or beetle grub, depending on the tachinid species, the maggot attacks the future host and bores through its skin. As a variant on this mode of attack there are those species which lay eggs on the leaf surface of a plant. The leaves are eaten and the eggs swallowed by a caterpillar. The tachinid eggs hatch in the gut of the host and the maggots bore through the wall of the gut into the internal tissues of the host. The young, newly-hatched maggot is very hardy and capable of surviving without food for some time. There is, nevertheless, a lot of wastage of young larvae and many never find a host.

In species with this method of host finding, the females produce thousands of minute eggs. In some species the female seeks out the host, be it caterpillar, beetle grub or other host and, with great care, deposits an egg on its body (Fig. 8). The egg hatches and the emerging larva bores through the skin of the host. Species which do this produce far fewer eggs than those which have less chance of success. Even so, there are dangers and great losses to the parasitic population. If it takes time, say a

Plate 37. Hover Fly, (Fam. Syrphidae). Egg, x15 **top left;** maggot, x8 **top right;** pupa, x5 **bottom left. Bottom right:** Drone Fly, *Eristalis tenax*. (Fam. Syrphidae). x3.

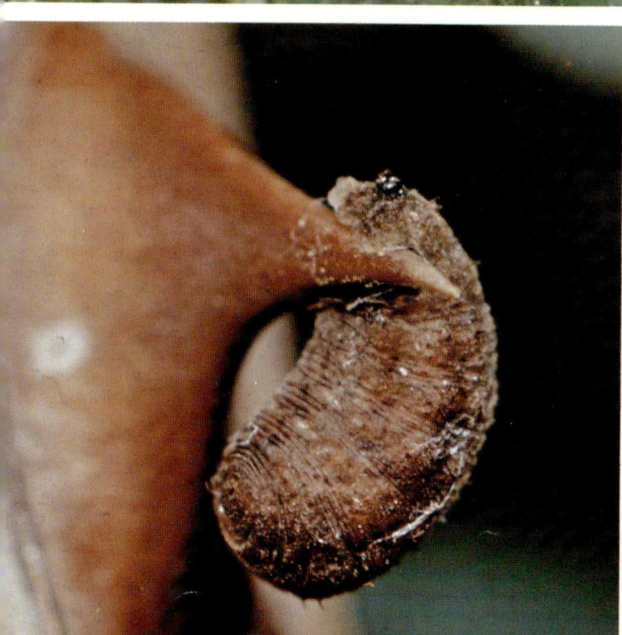

few days, for the egg to hatch, the host may moult and so deprive the maggot of a host, the egg being cast off with the old skin. In other species the female fly retains the egg in her body, and does not lay it until it is ready to hatch. She deposits a maggot instead of an egg. Using this system a female has to have the ability to find suitable hosts at the time the larva is ready to be deposited. In the types mentioned so far the parasite is deposited on the host and then has to bore its way into the host body to reach the internal tissues.

One additional refinement has been evolved by some species. The female is equipped with a sharp, curved, scimitar-like organ at the end of her abdomen. With this she makes a slashing action at the host, usually as she walks or flies over it; the skin is cut and an egg or maggot deposited directly into the host body. The action of the female may be so rapid that it is not easy to see what has happened and the host may give no more than a twitch of its body in response to an action which will lead to its death.

The parasite maggot feeds on the body tissues of its host, although the host may continue to live an apparently normal life. It will feed, grow, moult and even, in some species, pupate, with the parasite inside continuing to feed and grow itself. Eventually, however, the parasite destroys the host tissues to such an extent that it dies. It is usual for the parasite maggot to make its way out of the host and enter the soil to pupate. A great many insect species are attacked by tachinid parasites. As many of the world's major pests are attacked by tachinids, their importance to man is obvious.

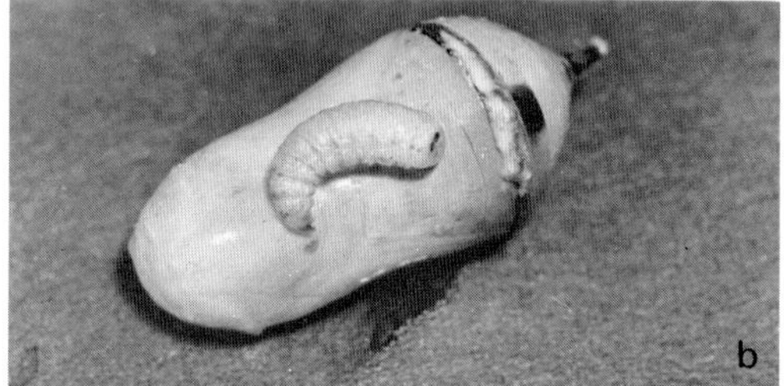

Fig. 8. Life cycle of parasitic Tachinid fly, *Winthemia diversa*. **a** Eggs on caterpillar. x5. **b** Maggot after emergence from host pupa. x1½. **c** Fly pupae. x1½. **d** Adult fly. x2½.

Plate 38. Tachinid Fly, *Microtropesa sinuata*. (Fam. Tachinidae). x6. 

FLEAS
(Order Siphonaptera)

TO THE DOG or cat-owner, fleas are a nuisance and sometimes a source of discomfort. On the other hand, they are quite interesting insects.

If you examine a flea closely (an ordinary magnifying glass will show you quite a lot) you will notice that it is typically 'insect', but that it has certain peculiarities of its own (Pl. 39). It has a very narrow compressed body; it has rows of hairs and spines which project backwards, while the antennae, conspicuous in so many insects, are very small and fold away neatly into special grooves. These features in the design of the insect are not there merely by chance, they are all features which make it fitted for life in a forest of hair. As the flea spends much of its time on an active host animal, it needs, and has, relatively large claws which enable it to cling to its host.

In the distant past, the ancestors of modern fleas possessed wings and were capable of flight. Wings may well be of great benefit out in the open but in the confined spaces of a heavy fur they would be a hindrance. So the flea has lost its wings. The problem of getting about and finding suitable hosts on which to feed remains, however, and here lies one of the reasons for the flea's leaping abilities. At the same time it has developed a remarkable sensitivity to the presence of suitable hosts in its vicinity. You have only to enter a flea-infested room which has not been occupied for some time to see how quickly your presence is detected and how soon they find their way to you seeking a meal.

Fleas feed on blood sucked from their hosts. Their mouthparts are specially adapted for this and consist of a set of tiny blades and channels which are used for cutting through skin and for sucking up blood. The flea lays its eggs, which are small, oval, white or creamy-white objects, loosely amongst the hairs of its host. The eggs eventually fall from the fur onto the ground. They often accumulate in places where flea infested animals sleep. The small white larva which emerges from the egg is not a blood feeder but feeds on pieces of skin, straw and particles of dirt and dust of animal origin. The young active flea larva stands a fair chance, therefore, of hatching in a place in which it will not be too difficult to find food.

When fully grown the larva spins a small silken cocoon in some crevice near its feeding place. Within the cocoon it becomes a pupa, from which the adult flea eventually emerges. The whole cycle takes anything from three to eleven weeks, according to the species of flea and the temperature conditions at the time. It is known that slight vibrations may make the adult fleas emerge from the cocoons; this means that the fleas can lie protected within the cocoons for some time and as soon as vibrations occur, such as those produced by a likely host moving about nearby, they can emerge and be assured of a fair chance of getting a meal. This is another case of the remarkable sensitivity to outside influences which these insects have developed.

Fleas are usually regarded as repulsive by human beings; this is largely due to the fact that they are parasites. Nevertheless, they have a place in the general scheme of things and have evolved into creatures adapted to a fine degree in special ways for their mode of life. Many species are restricted to one or a few species of host; some fairly large species are found on marsupials in Australia.

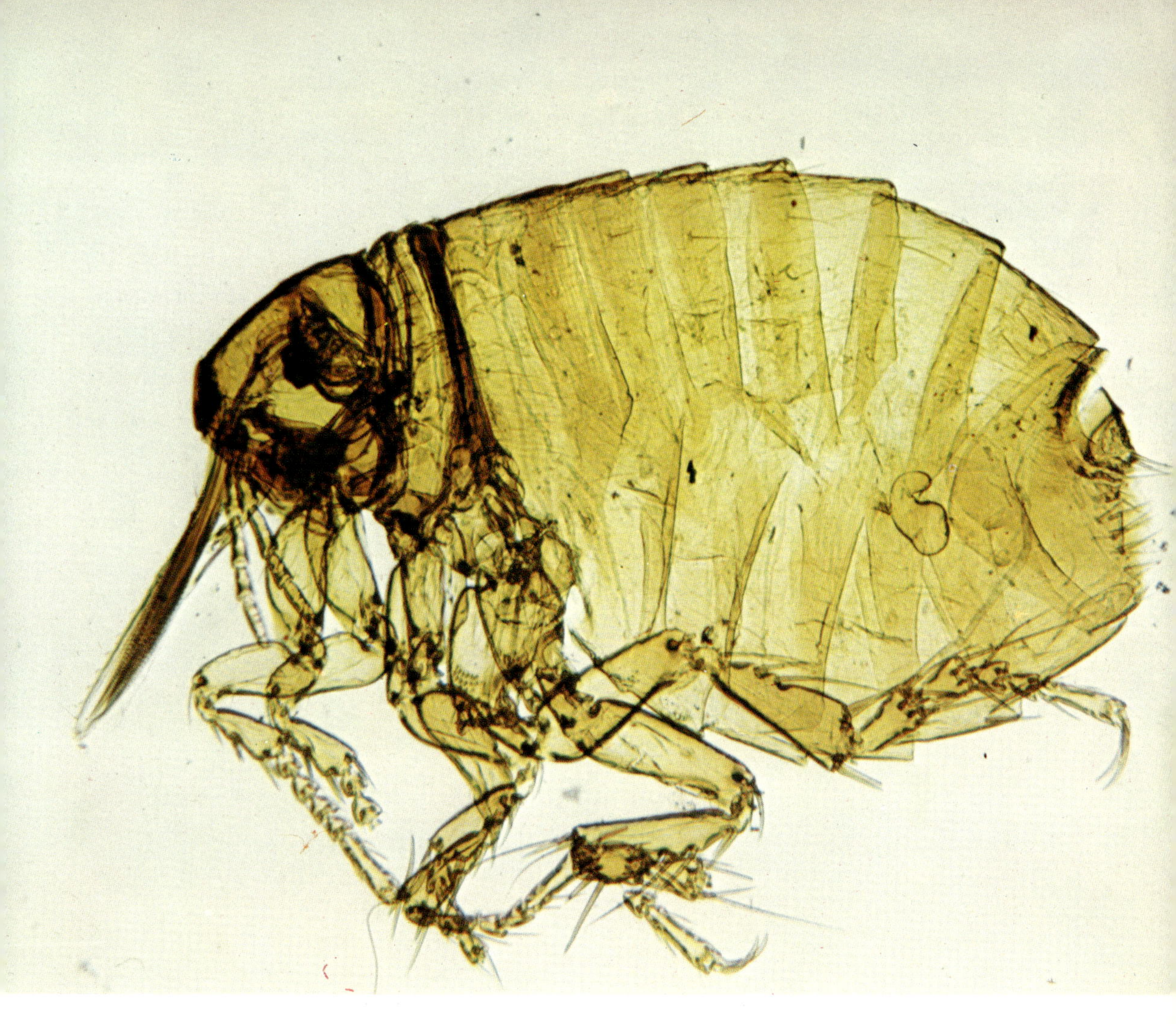

Plate 39. Flea, *Echidnophaga ambulans*. (Fam. Pulicidae). On microscope slide, cleared. x75.

ANTS, BEES, WASPS
(Order Hymenoptera)

THIS GROUP OF INSECTS includes the ants, bees, wasps, ichneumonflies and a vast array of other species which do not have common names. The degree of specialisation which many species have achieved, not only in the evolution of their physical characters, but also in their complex behaviour patterns, exceeds that of any other order.

The ways of life to be found in this order are remarkably varied; some groups are foliage feeders in their larval stage. Many are parasitic when immature. In other groups, all stages in the development of social behaviour leading to formation of the complex colonies of the truly social insects, can be seen.

Fig. 9. Cluster of Sawfly larvae, *Perga* sp. x¾.

SAWFLIES—*Suborder Symphyta*

The sawflies (Pl. 40, Fig. 9) are the most primitive of the Hymenoptera; they have features in both structure and biology which they have retained from their predecessors. They do not have the strongly constricted 'wasp-waist' so characteristic of ants, wasps and bees, and they have remained vegetarians in the larval state. Australia has some particularly interesting species of sawflies.

In most species the female has a saw-like structure for sawing into plant material and making slits, into which she deposits her eggs. The larvae of sawflies have a general resemblance to caterpillars, with a well developed head and long body with legs on the thorax and prolegs, similar to those of caterpillars, on the abdomen. There are wood boring and leaf mining species and here the larvae have lost the legs so that they are slug-like in appearance; this is an adaptation to their mode of living internally in plant tissues. The leaf-chewing species are capable of eating a considerable volume of leaves during their development and trees may be completely defoliated. Some of the leaf feeders are quite large insects and some species have the habit of clustering together on the trees (Fig. 9).

Plate 40. Top: Sawfly, *Platypsectra analis.* (Fam. Pergidae). x4¼. **Bottom left:** Pear and Cherry Slug, Sawfly larva, *Caliroa cerasi.* (Fam. Tenthredinidae). x16. **Bottom right:** Sawfly larva, *Platysectra analis.* x7½.

In Australia these are commonly called 'spitfires'. When feeding the larvae disperse but come together again later. These masses of larvae are conspicuous particularly as they become older and larger. When they have finished feeding and are nearing pupation they leave the trees and move into the soil, usually near the trees on which they have been feeding. Each one spins a silken cocoon in which to pupate and it is possible to find masses of these pupae below trees which have been heavily attacked.

Suborder Apocrita

All other members of the Hymenoptera have been grouped by entomologists under the suborder name Apocrita.

ICHNEUMONIDS—*Family Ichneumonidae*

Ichneumonids (front cover, Pl. 41) are usually slender, wasp-like insects, with long bodies. The life history of most ichneumonids is somewhat bizarre and is a type of life cycle shared with other insects of the order Hymenoptera as well as many of the parasitic flies, such as the Tachinidae. The female seeks out a suitable host insect. This varies according to the species of ichneumonid, some requiring a very definite host species, others being satisfied with any one of a large range of species. Most species attack caterpillars but the larvae of other orders, such as Coleoptera and Diptera, are also used as hosts. The female ichneumonid thrusts her ovipositor into the host's body. This frequently has the effect of at least partly paralysing it. She then lays an egg. The larva which hatches from the parasite egg is a colourless, legless grub which, being immersed in the highly nutritious tissues of its host, does not need to be particularly mobile. The parasite feeds on the tissues of the host, eventually killing it.

In some species the ichneumonid lays her egg externally. The grub may enter the host's body and become an internal parasite or may attack the host from the outside. When the grub is fully developed it usually leaves the body of the host, or what is left of it, and spins a cocoon in which to pupate. The number of host specimens annually accounted for in this way by ichneumonids and other parasitic wasps is beyond calculation. When caterpillars, such as army worms, reach plague proportions and are busy devastating pastures or crops, it is not unusual to see thousands of ichneumonids over the vegetation seeking hosts. The pest has probably done considerable damage by this time, but the influence of the ichneumonids in preventing many many of them from reaching maturity, and so reproducing, is great.

BRACONIDS—*Family Braconidae*

The braconids (Pl. 42) are very similar to the ichneumonids and are separated from them by entomologists on the basis of certain anatomical differences, especially of the wing veins. In general, they have similar types of life cycle and also attack a wide range of host species. The most commonly seen species are those of the genus *Apanteles* which attack caterpillars such as those of the cabbage white butterfly. The adult braconids of this genus are small and as many as a hundred or more are able to develop in one caterpillar. The grubs are small and have the habit of all emerging from the remains of the host soon after one another or even more or less simultaneously. The grubs do not move far from the host body before pupating, each spinning a small cocoon, often of golden-coloured silk. The cocoons are formed close to one another, often around the host (Pl. 42). Such groups of cocoons are often found in fields of cabbages or cauliflowers which have been heavily attacked by cabbage white caterpillars but the caterpillars on many other plants may be similarly attacked.

Plate 41. Ichneumonid Wasp. (Fam. Ichneumonidae). x7.

Even smaller species of braconids attack aphids. If a strong colony of aphids on a bean or rose shoot is examined it is usually possible to find amongst the living aphids the remains of at least some specimens (sometimes a high proportion of specimens) which are brownish or buff-coloured. These are parasitised individuals which have died from the internal attacks of a braconid grub. The grub, in this case, has pupated within the remains of the host and there is often a circular hole, caused by the adult wasp as it emerged from the host body.

WASPS—*Family Vespidae*

Although the most familiar wasps live in communal nests, a great many species of wasps are not social insects, but live a solitary existence. They burrow in soil or wood and make a series of cells which are stocked with insects or spiders of various kinds, often caterpillars, which have been immobilised by stinging. Some species make cells of mud. An egg is placed in each cell and the cell closed up. The wasp larva feeds on the store of food and eventually emerges as a wasp after passing through its pupal stage. Wasps are very common in tropical and warm temperate areas and when their populations are large can be responsible for considerable reduction in numbers of their prey species.

Wasps, too, have evolved a form of social life and although it parallels that of the ants it has not reached quite the same levels of specialisation.

Reference has already been made to the social arrangements and different types of individuals to be found in a termite colony. In wasps we again find a complex society but with certain basic differences. It is clear that the phenomenon which we call 'social life' has arisen more than once in the evolutionary history of the insects, another example of the ability of this group of animals to exploit the environment efficiently. Social life is more than a mere gathering together. When wanderer butterflies gather together in winter or when the larvae of saw-flies cluster when they are not feeding the whole thing is a rather temporary affair. In the case of social insects it seems that we need to look for some other kind of beginning; a mere gathering together of like individuals is not enough to provide the beginnings of a social structure. If we look at the behaviour of some of the species of earwigs or some of the sucking bugs, we find that the female parent stays with her eggs and young for some time. One of the main differences between social insects and others is that in the social species the female parent is assisted in some way by her off-spring to care for the subsequent generation. In order to achieve this the female must live long enough to be able to associate with her offspring. This we find in such insects as the earwigs; they have the kind of parent-offspring overlap in time and continued association which could conceivably lead to the evolution of a society of some sort. A social insect can be defined as one in which parent and offspring associate in mutual cooperation in a common nest or shelter.

In a colony of *Polistes* (Pl. 43) or *Ropalidia* (Fig. 10) the nest is made of 'paper'; this is a material derived from chewed up wood and secretions from the wasp. In *Polistes* it is suspended, usually in a protected situation, and consists of a layer of cells forming a structure analogous to the 'comb' of the honey-bee. An egg is laid in each of the cells and

Plate 42. **Top:** Braconid Wasp, *Bracon capitator*. (Fam. Braconidae). x5. **Bottom:** Cocoons of Braconid wasp, *Apanteles* sp. (Fam. Braconidae) with host species of caterpillar, Cabbage White Butterfly, *Pieris rapae*. (Fam. Pieridae). x6.

the larvae are fed on masticated caterpillars provided progressively by adult wasps. Before the larvae pupate they spin a silken seal to the cell. There is no great difference in body form between members of the colony but egg laying is, in the main, restricted to a queen whilst the collection of food and building material and the care of larvae are the responsibilities of sterile workers, which are all females.

As the season progresses, males and fertile females are produced, these mate and the fertile females over-winter, emerging the following season to start the construction of small new nests; the males do not survive the winter in cooler areas. A remarkable relationship arises in social wasps between the workers and the larvae they feed and 'care for'. The larvae produce a secretion which is much sought after by the workers and which is produced in response to manipulation of the larvae by the mouthparts of the worker. In return, as it were, the larvae are fed; their neglect would lead to a cessation of the flow of the secretions. At times when food is scarce the larvae may be so harassed by the workers seeking the secretions that they do not grow normally. The social bond which ensures stability of the colony is dependent on this mutual feeding.

Although the colonies of wasps remain fairly small in temperate regions, they are large in some tropical Queensland species and the effects of combined attack by workers on intruders can be dangerous.

ANTS—*Family Formicidae*

Ants (Pls. 44, 45, Fig. 11) have evolved the most advanced forms of social life to be found amongst the insects. The bull ants (Pl. 44, Fig. 11) of Australia are amongst the most primitive; they have relatively small and simple colonies and they have retained a very well developed and functional sting in the worker caste. Ants have a much more varied diet than the other social insects; they are able to exploit a wider range of foods. They attack other insects, feed on seeds, fungi, meat and on the sweet secretions (honey-dew) of other insects. Foraging worker ants cover almost every square inch of ground and soon find any corpse; in a short time its existence has been made known to other members of the colony and it is being removed. Nearly all ant nests are subterranean and large quantities of earth are moved by the tunnelling activities of ants. By using the soil as a nesting site the ants spend less energy and can work more quickly than bees and wasps in nest formation; the nests can also be much more extensive and house large populations.

Owing to the fact that the nests are subterranean and most of their activities take place on the ground and on trees, ants have been able to dispense with wings, retaining them, like the termites in a similar situation, only for the nuptial flight. In the case of

Fig. 10. Six inch long nests of paper wasps in sandstone cave, *Ropalidia* sp.

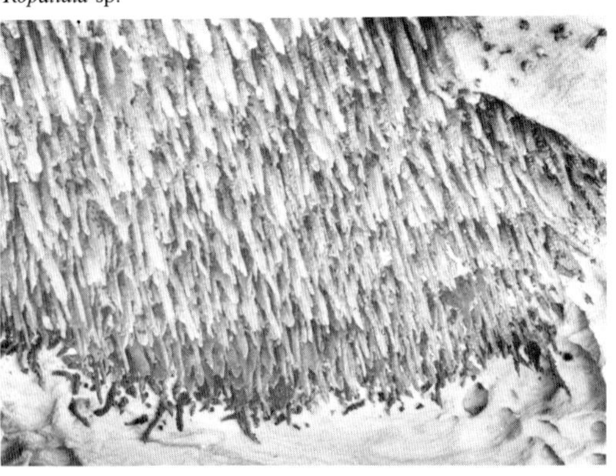

Plate 43. Paper Wasp, *Polistes humilis*. (Fam. Vespidae). x5¼.

a typical species, winged males and females (Pl. 45) emerge from the nests in great numbers over a brief period in spring or summer. These flights often involve enormous numbers of individuals. In many species the flights are often associated with certain meteorological conditions, hence the many accounts of ants being forecasters of certain weather patterns. After mating, the female comes to earth and loses her wings. She only mates once and this is sufficient to enable her to produce fertilised eggs for the rest of her life, which may be several years. Depending on the species, she may return to her old nest, she may enter another nest of the same species or she may start a new colony. Only a very small proportion of the winged individuals which leave a nest succeed in setting up a new colony; the swarms are attacked by a great many predators. Birds, insects, and reptiles all take heavy toll of the ants as they leave the nest or as they soar into the air on their one and only flight.

A female which has succeeded in mating enters a cavity in the soil or excavates one for herself. Here, usually after some time, she lays a few eggs. The larvae are reared on salivary secretions produced by the female and during this period food reserves and tissues of her own body which will no longer be needed, such as wing muscles, are broken down to provide food for her offspring. These early larvae always become sterile wingless female workers (Pl. 45), and they take over from the original female (queen) the tasks of running the nest, other than that of egg-laying. They forage for food, defend the nest and take care of the queen and larvae, cleaning and feeding the larvae and moving them from one part of the nest to another as conditions of moisture and temperature change. They keep the eggs, larvae and pupae in separate chambers of the nest, at the same time extending the tunnels and chambers to meet the needs of the developing colony. As in the case of the wasps, the larvae are 'licked' and provide secretions which are taken up by the workers. This exchange of food is the main bond which preserves the colony as an entity.

Fig. 11. Head of Bull Ant, *Myrmecia nigriceps*. x6½.

The speed with which ants will detect and exploit a food source is often quite remarkable. This is achieved by having foraging individuals which, when they come upon a food source, return to the nest leaving a scent trail (the substance released being technically known as a 'pheromone') which can be followed out from the nest to the food by other workers. In this way, other workers can find the food source. In other cases continuous feeding at an established source, such as a colony of aphids producing honey-dew, is possible. Ant colonies can reach incredible numbers, 100,000 or more is not uncommon in some parts of the world. Very little ground or the vegetation growing on it is not within the normal foraging area of one or other ant colony. Competition between colonies is strong. Many animals, naturally enough, have taken to

Plate 44. Bull Ant, *Myrmecia nigriceps*. (Fam. Formicidae). x8.

using this supply of insects as a food source and ants are preyed upon by many species of birds, insects and reptiles as continuous food sources.

The main enemies of ants, however, are other ants! Some species prey on others, attacking their nests and carrying off eggs, larvae and pupae for food. Some species have developed special castes of sterile females which do nothing but defend the colony. These 'soldiers' are usually more heavily built than the workers and have stronger jaws and often bigger heads. Their specialisation as colony defenders is sometimes carried to such extremes that they cannot feed themselves but are fed by workers. This development of special soldiers is yet another case of parallel evolution between ants and termites, two groups of insects only distantly related but which have both evolved complex societies. They have met and solved the same problems of survival in similar ways. The other great social animal, man, also has problems of survival; it remains to be seen whether he can solve them.

Humidity and temperature in the nests are held remarkably constant by the activities of the insects themselves. With such a stable environment available it is not surprising that other species of insects, in the manner so typical of such an enterprising group of animals, have invaded the nests and taken up residence. Within the nests we find a great variety of insects other than ants living in the galleries and chambers with them. These include various species of beetles, flies, springtails and sucking bugs. Some of them live amicably with the ants, some ignore the ants, others are scavengers which live on the debris within the nest or are predators, feeding on the ants, their eggs and the grubs.

With such activity going on within a single nest, it is not surprising that ants are amongst the most fascinating of insects, nor that they have attracted the attention of some of the world's most notable entomologists.

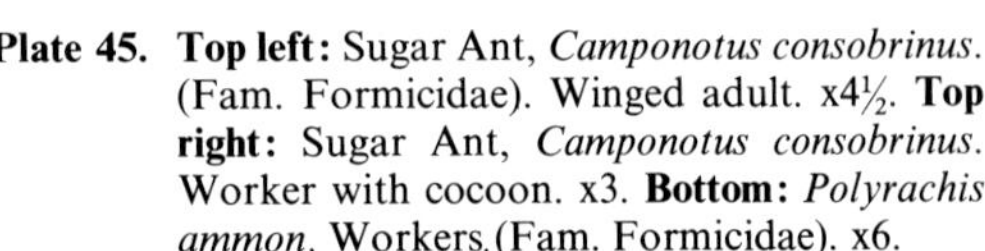

Plate 45. **Top left:** Sugar Ant, *Camponotus consobrinus*. (Fam. Formicidae). Winged adult. x4½. **Top right:** Sugar Ant, *Camponotus consobrinus*. Worker with cocoon. x3. **Bottom:** *Polyrachis ammon,* Workers.(Fam. Formicidae). x6.

BEETLES

(Order Coleoptera)

THE BEETLES (Pls. 46–52) make up the biggest order of animals; there are over 200,000 species known in the world, of which something less than 20,000 are Australian. Each year many more species are discovered and described. Despite these figures, it would be true to say that beetles as a group are not by any means the most conspicuous insects, although many are quite large. With the large numbers of species in existence it is only to be expected that they should have exploited almost every available habitat, but they tend to remain concealed. In becoming adapted to a wide range of habitats they have also evolved widely differing habits. Many of them are inhabitants of trees and shrubs, especially as adults. Several times in their evolutionary history they have invaded freshwater to a greater or lesser extent but have retained the ability to fly from one body of water to another. They are found in carcasses, in fungi, and in all types of rotting animal and vegetable matter. Several species are closely associated with the products which man stores in bulk, such as grain, skins, timber and processed foods. Above all, however, they have colonised the soil and live on or in it, feeding on vegetable and animal materials. For this reason they tend not to be noticed. Turn over any log or stone in a forest or disturb any leaf litter or debris lying on the ground and you are very likely to find a beetle of some sort.

Of all orders of insects, the beetles are the most easily recognised because the front pair of wings has become thickened and hardened to form structures which serve as covers, called elytra (singular—elytron) for the more delicate, functional, pair of hind wings. In some families the elytra have become shortened so that the end of the body is exposed. This is so, for example, in the very large family of rove beetles (Staphylinidae) (Pl. 47). Their elytra are so shortened that they sometimes cover only a quarter or less of the abdomen. This gives the beetles a rather earwig-like appearance but rove beetles do not have pincers at the end of the abdomen. In the process of becoming hardened the elytra have, in many cases, lost their wing-like shape and they are not used for flight. When the insect flies the elytra are usually held up and away from the body leaving room for the hind wings to beat. As soon as the insect settles, the hind wings are folded and tucked away, sometimes in quite a complex way.

The upper surface of the body of beetles is usually hard and this hardness has probably been responsible for their protection against a wide variety of enemies. Most beetles are of generally sombre colour, with browns and black predominating but there are colourful exceptions, especially in species which live amongst green vegetation or are associated with flowers. On the whole, however, ground-dwelling insects do not assume bright colour patterns unless they are distasteful to predators or are particularly pugnacious and so can afford to advertise their presence. Many beetles are nocturnal, resting by day and coming out after dark to feed. Beetles have always been favourites for col-

Plate 46. Top left: Chrysomelid Beetle, *Brachychaulus klugi*. (Fam. Chrysomelidae). x6½. **Top right:** Lycid Beetle, *Metriorrhynchus rhipidius*. (Fam. Lycidae). x3¼. **Bottom left:** Fungus Beetle, *Episcaphula pictipennis*. (Fam. Erotylidae). x6½. **Bottom right:** Cantharid Beetle, *Cantharis curvipes*. (Fam. Cantharidae). x6½.

lectors, especially the larger species and those with colourful body patterns; some of these are now becoming rare because of the removal of their natural habitats. Some may already be extinct.

GROUND BEETLES—*Family Carabidae*

The beetles of this family, which is a large one, are nearly all predacious and as their common name implies, most of them live on or in the soil. Here they feed on other insects, centipedes, worms, slaters and other soil inhabiting animals. Most of them are rapid runners and have slender legs well suited for rapid movements. They usually have long antennae and powerful jaws and their bodies are hard and strong, all characters useful to an animal which lives by attacking and feeding on others, many of which are themselves active predators. Ground beetles are built for hunting. The larvae of ground beetles are also predatory creatures and they, too, are hard bodied, elongate and have well developed legs and jaws. Ground beetles are nearly all nocturnal, another reason why they are seldom seen unless specially searched for, although their populations are sometimes quite high. During the day they can be found under logs or stones and they usually hurry away into a crevice as soon as their cover is removed. Some of the more conspicuously marked species are capable of producing a liquid with caustic properties. This is squirted out from special organs into which the liquid is secreted by glands and when the insect is roughly handled or disturbed and irritated it is used in defence. In the case of the bombardier beetle liquid is squirted out from the posterior of the abdomen. When this comes into contact with the air it volatilises and appears as a small puff of 'smoke'. At the same time its extrusion gives rise to an audible popping sound. If the liquid comes in contact with the skin it causes an effect somewhat like that of a drop of acid; this is quite painful and may remain so for a while.

TRUE WATER BEETLES—*Family Dytiscidae*

Like other fast moving aquatic insects the water beetles (Pl. 47) have become stream-lined and smooth; they have reduced antennae and large, flattened, paddle-like hind legs. From time to time the beetle rises to the surface of the water to replenish its oxygen supply. It does this by protruding the end of its abdomen out of the water and replacing the air in the system of tracheae. The back of the abdomen, hidden from view by the elytra, is covered with fine hairs which form a thick pile. This retains a bubble of air which can also be replenished on each visit to the surface by raising the elytra. Water beetles feed on a wide variety of water creatures. Their larvae are long and thin and also breathe by coming to the surface from time to time and exposing their terminal spiracles. They do not, however, carry an additional reserve. They feed by means of their long mandibles, which are strongly grooved and sickle-shaped; they suck out the body juices of their prey through the almost closed groove. The adult beetles often fly at night and their flights may be quite long. Water beetles are readily attracted to light and occasionally arrive in large numbers a long way from water. By means of such nocturnal flights new bodies of water are colonised. Their appearance in swimming pools is sometimes of concern to the owners, but they do no harm and would only remove a few of the other unwanted inhabitants. In any case, they can seldom

Plate 47. Top: Rove Beetle, *Eulissus chalcopterus*. (Fam. Staphylinidae). x7¼. **Bottom:** Water Beetle, *Eretes australis*. (Fam. Dytiscidae). x5.

breed and mature under swimming pool conditions. Some species are known which are normally inhabitants of brackish or salt water and a few have been found living in thermal pools.

WHIRLIGIG BEETLES—*Family Gyrinidae*

The whirligig beetles (Fig. 12) are well known to anyone who has lazed away time on the bank of a quiet stream. They are oval-bodied beetles and spend much of their day whirling around each other in graceful curves in the water surface. They send out ripples as they swim and their movements become difficult to follow. They are seldom alone for long, when one begins its gyrations it is soon joined by others. Whirligigs move in the surface, that is, part of their body is in and part out of the water. Their eyes are peculiar in that they are divided into an upper and lower half, probably as an adaptation to vision below and above the water. The beetle's underside is smooth and stream-lined; its middle and hind legs are reduced in size but are wide and flattened and have fringes of hairs which increase their area. They cannot be seen from above but when the beetle is swimming they are paddling rapidly below the water surface. The front legs, unlike those of the water beetles, are long and are used for grabbing at the prey on which the beetles feed. Their prey consists of insects on or trapped in the water surface; the adult beetles are nearly always predacious. Their larvae are long and narrow and live near the bottom of the stream or pond. They do not need to make the periodic visits to the surface which are so vital to true water beetle larvae; they have finely branched, delicate, feathery structures arising from the sides of each segment of the abdomen, which function as gills, capable of extracting oxygen from the water and expelling the carbon dioxide produced as a result of the insects' respiratory activities. Like the water beetles, whirligigs occasionally set off in search of new homes and may arrive without warning in swimming pools.

JEWEL BEETLES—*Family Buprestidae*

The jewel beetles (Pl. 48) as their name would suggest, are amongst the brightest and most colourful of the beetles. Some species are also quite large. The colours include reds, yellows and beautiful, iridescent, metallic colours. They are elongate beetles, tending to taper towards the hind end. They are most easily located on trees and shrubs in flower. They are active flyers and usually take off on hot summer days, making circling flights in search of flowering plants. The larvae are mainly found in wood, either in the stems or roots and

Fig. 12. Whirligig beetles, *Macrogyrus oblongus*. x2½.

Plate 48. Jewel Beetle, *Melobasis cuprifera*. (Fam. Buprestidae). x4½.

less often as borers in herbaceous plant stems. In some species the larvae appear to be remarkably long-lived as the adults have emerged from the timber of houses and from pieces of furniture, made of local woods, many years after construction.

FIREFLIES—*Family Lampyridae*

The flashing lights of fireflies is one of the delights of warm tropical forests at night. Each flash gives the appearance, to our eyes, of a streak of colour. The frequency of the flashing, length of the flash and flight path, go to make up a flash pattern which is distinctive for some species. Sometimes the larvae, pupae and even eggs are capable of producing light. The flashing serves to bring the two sexes together, but it is difficult to decide on its significance, if any, for the immature stages. Both sexes are usually luminous but the females, although they have wings, have not been seen flying. The adults do not seem to feed although the larvae are voracious feeders and attack mainly snails. They inject paralysing fluid into the prey and also juices which digest its tissues. The resultant nutritious fluid is re-imbibed by the larvae.

The light produced by Australian species of fireflies is a pale greenish colour. The light producing tissues are situated on the underside of the abdomen where the integument is transparent. Below this is another layer which reflects the light outwards. The light producing organs are well supplied with oxygen and light production is carried out by the transfer of oxygen to a substance called luciferin, in the presence of another substance called luciferase. The precise chemical nature of the substance receiving the oxygen determines the colour of light produced. Flashing may be controlled through the nervous system or by control of the supply of oxygen or by the availability of the material to be oxidised. At present we are ignorant on these things. No matter how little we might know of the light produced by fireflies, we can still admire their glowing flashes in the forests.

LADYBIRDS—*Family Coccinellidae*

Most of us meet a ladybird (Pl. 49) at an early age. Although it was probably not realised when the tradition started, the image of the ladybird as a friend is certainly justified. Some species are amongst the most efficient predators of scale insects, aphids, mites and other small pest species. A few species do feed on vegetation and are pests themselves; these usually attack the leaves of potatoes and pumpkins. The predacious species are voracious feeders in both larval and adult stages and can destroy large numbers of prey. Their rate of reproduction is fairly high. The adults lay their golden eggs in groups, usually on plants on which numbers of prey are available. These hatch and the larvae soon set off in search of prey. If well supplied with food their development is rapid and they pupate in a matter of days. The population of ladybirds grows rapidly under favourable conditions.

Ladybirds seem to be distasteful to predators; if they are disturbed they are capable of deliberately exuding their poisonous blood from a point at one of the leg joints, a process known as reflex bleeding. The conspicuous colours of most species are probably a warning to would-be attackers. Like some species of butterflies, some still unmated ladybirds migrate at the end of the breeding season and cluster together in large numbers, usually in some secluded position such as under bark or in build-

Plate 49. Ladybird Beetle, *Leis conformis*. (Fam. Coccinellidae). Larva, x22 **top;** pupa, x7½ **bottom left;** adult, x9 **bottom right.**

ings. At the end of this hibernation period they emerge and mate and set off again to areas where food supplies are available. In most cases the places chosen for clustering are on a hilltop or other point high above the surrounding country.

CHAFERS, CHRISTMAS BEETLES, SCARABS—*Family Scarabaeidae*

The 'scarab' beetles (Pls. 50, 51) are probably amongst the most familiar of beetles. They constitute a large family, with over two thousand species in Australia. They are often large or moderately large, stout insects in which the body is convex. In some species bright colours, mainly iridescent, or bold patterns, make the beetles quite handsome. In others, protuberances on the head and body occur of complex and strange shapes, giving the animal a grotesque appearance; these are usually larger and more complex in the males than in the female.

Many species have a close association with the droppings of animals. Some have the habit of dung-rolling, in which a piece of dung is made into a ball and rolled along by the beetle, sometimes for a long distance. The ball of dung is then buried. It may be eaten by the adult beetle or an egg deposited on it, in which case the resulting larva will use the ball as a food source. This habit parallels that of some wasps and bees which store cells of food for their larvae. The importance of dung beetles as scavengers which remove animal droppings is considerable under natural conditions in areas where there are large populations of animals.

On the other hand, there are many species which are serious pests where the adults are leaf feeders and, when populations are high, are capable of defoliating trees. The larvae (Fig. 13) of members of this family, known as 'white grubs' or 'curl grubs', are very familiar to gardeners. They are white, with reduced legs and with a hard brownish head. The body is thick, particularly towards the hind end and is curved. Many species occur in areas where the soil has high humus content and in compost heaps; others are found in pastures and lawns. They feed on rotting vegetable material or on roots. When populations are high in grassy areas, considerable damage may be done by the grubs severing roots. Brown patches appear where the grass dies back and dead turf can be lifted up in sheets. In addition to the pests of pastures, there are other species which are pests of crops such as maize and sugar cane. The adults of most of the pest species are nocturnal and dull-coloured and some of them are readily attracted to light. Some are active for only a short period in spring or

Fig. 13. White Grub or Curl Grub, larva of scarab beetle. x2.

Plate 50. Fiddler Beetle, *Eupoecila australasiae*. (Fam. Scarabaeidae). x5½.

Fig. 14. Head of Long-horned beetle, *Eurynassa* sp. x4.

summer when there is a massed emergence of adults. At this time enormous numbers may invade homes at night; they are a nuisance but do no damage indoors.

The brightly coloured species are mainly diurnal; many of these are found on flowers, where they feed on nectar or pollen; a few feed at ripe or over-ripe fruit.

Fig. 15. Larva of Long-horned beetle, *Eurynassa* sp. x1½.

LONG-HORNED BEETLES—
Family Cerambycidae

The long-horned beetles (Figs. 14, 15) are usually robust beetles, with somewhat cylindrical bodies and well developed, conspicuous antennae and stout mandibles. In the larval stages they are wood borers (Fig. 15); whilst most species attack dead wood or dying trees a few will feed in living plants. The female lays her eggs in a crevice in the bark and the larvae tunnel into the wood on which they feed.

All wood feeding insects face a physiological problem in that the constituents of wood are notoriously indigestible. The termites overcome the problem of digesting cellulose, the major constituent of cell walls of plants, by maintaining within their digestive systems a culture of micro-organsims which are capable of breaking it down. The larvae of long-horned beetles, however, are amongst the few insects capable of digesting cellulose themselves. This they do with a special secretion which is capable of breaking the cellulose into simpler substances which can be absorbed through the wall of the gut into the blood. Wood is not a particularly nutritious material, in any case, and a relatively large quantity has to be taken in and digested in order to extract sufficient nutriment to maintain a reasonable rate of development.

Like some of the jewel beetles many of the long-horned beetles may take several years to complete their life cycle. This applies, of course, particularly to the larger species. Cases are also known of long-horned beetles emerging from furniture and other

Plate 51. **Top:** Belid Weevil, *Belus filiformis.* (Fam. Curculionidae). x6. **Bottom left:** Weevil, *Leptopius duponti.* (Fam. Curculionidae). x3½. **Bottom right:** Scarab beetles, *Diphucephala pigmaea.* (Fam. Scarabaeidae). x6.

timber years after it has been made up and in use. Larvae sometimes occur in large numbers in fallen logs and the sounds of their chewing and moving in their tunnels can be heard from several feet away.

LEAF BEETLES—*Family Chrysomelidae*

The leaf beetles (Pl. 46) are seldom large, but there are many extremely beautiful species; in many cases the patterns include bright, iridescent, metallic colours of exceptional brilliance. Although the vast majority of species in this very large family are terrestrial in all stages, a few have aquatic larvae. Species of *Paropsis* are common on *Eucalyptus* in Australia and both adults and larvae feed extensively on the leaves, sometimes causing defoliation. They lay their eggs (Fig. 16) in groups which form bands around shoots on the host plants. The larvae have the remarkable ability to produce highly poisonous hydrocyanic acid. A small proportion of species feeds on crops but the majority feed on native plants and their populations usually remain small.

Fig. 16. Chrysomelid beetle eggs, *Paropsis* sp. x3.

WEEVILS—*Family Curculionidae*

This family is the largest in the animal kingdom with about 60,000 species known in the world, of which at least 4,000 are Australian. It contains all the beetles commonly known as weevils (Pls. 51, 52). They are characterised by having the head drawn out in front into a more or less strong snout at the end of which are situated the mouthparts. The antennae arise on either side near the base of the snout and are bent in the middle, so having an elbowed appearance. Like most insect structures the snout varies in degree of development from species to species; in some it is only a short, stout extension of the head, at the other extreme it is in the form of a long, fine, needle-like prolongation. In some species there are differences in degree of snout development in males and females. The female in some cases uses the snout as a drilling instrument, making a cavity in plant tissue in which she later places her eggs. In many species, however, it does not seem to have this function. The eggs, laid in protected situations, are smooth shelled and colourless. There seems to be no particular use for the snout in the males. Whilst most weevils are dull-coloured, a few such as the Botany Bay diamond beetle (Pl. 52), one of the first insects to be collected in Australia, are brightly coloured. A large number of species have the body clothed with scales, as in the butterflies and moths; this is unusual for beetles. Weevil larvae are legless, grub-like creatures; most of them live underground where they attack the roots of plants or they live internally in plant tissue; a few live on aquatic plants. Some of the most injurious pest species are weevils. Members of this family are the worst pests of stores of human grain foods.

Plate 52. Botany Bay Diamond Beetle, *Chrysolophus spectabilis*. (Fam. Curculionidae). x8.

Index

Bold figures indicate colour plates or black and white figures

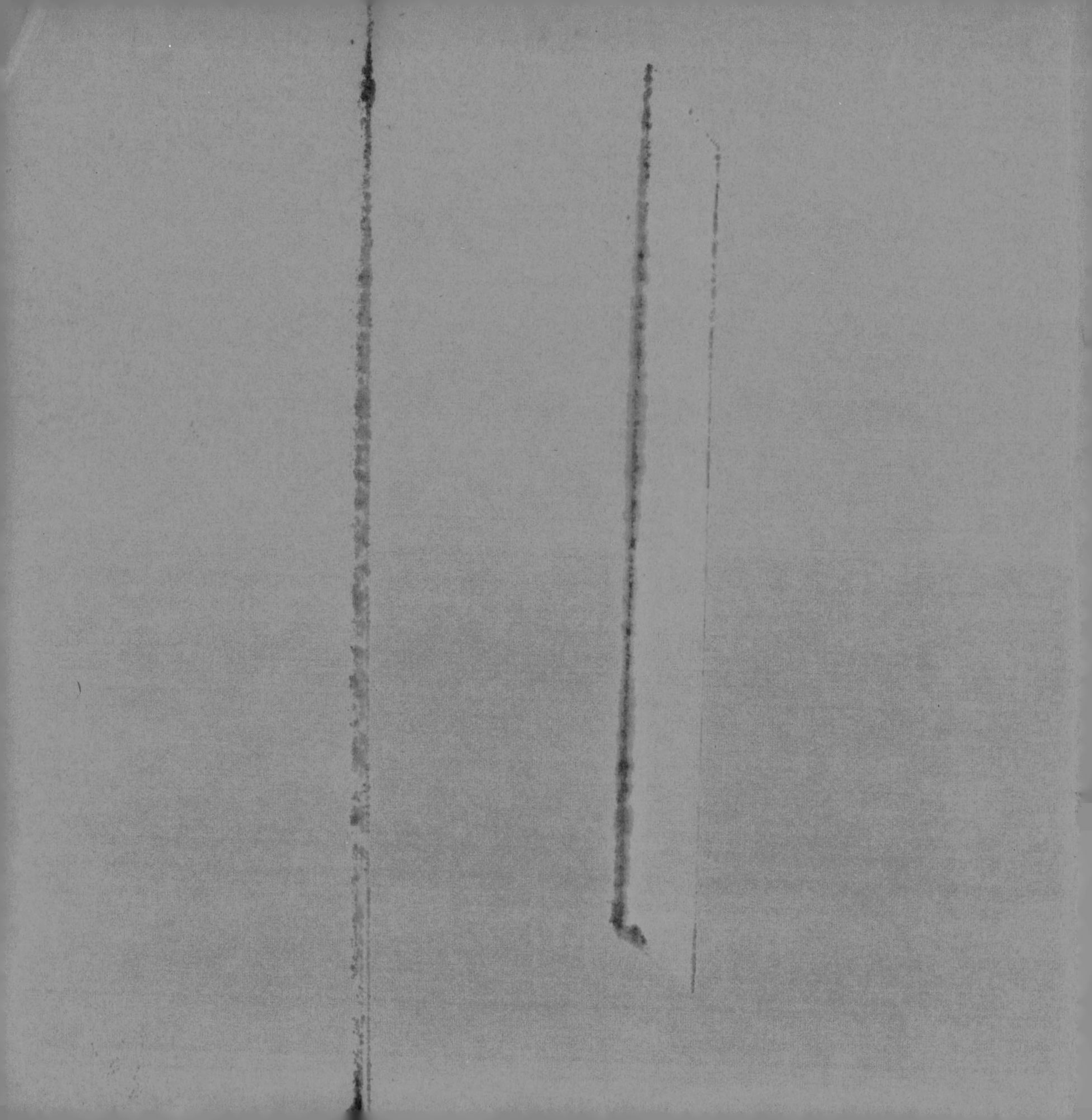